NORKAM SECONDARY SCHOOL
730 - 12th STREET
KAMLOOPS, BC V2B 3C1
PH (250) 376-1272
FAX (250) 376-3142

Date Due

JUN 1 2 2000			

BRODART. Cat. No. 23 233 Printed in U.S.A.

THE COMPLETE WEATHER RESOURCE

THE COMPLETE WEATHER RESOURCE

Volume 2: Weather Phenomena

Phillis Engelbert

AN IMPRINT OF GALE

DETROIT · NEW YORK · TORONTO · LONDON

THE COMPLETE WEATHER RESOURCE

by Phillis Engelbert

Staff

Jane Hoehner, *U•X•L Senior Editor*
Carol DeKane Nagel, *U•X•L Managing Editor*
Thomas L. Romig, *U•X•L Publisher*

Mary Beth Trimper, *Production Director*
Evi Seoud, *Assistant Production Manager*
Shanna P. Heilveil, *Production Assistant*

Cynthia Baldwin, *Product Design Manager*
Barbara J. Yarrow, *Graphic Services Supervisor*
Pamela A. E. Galbreath, *Cover and Page Designer*

Margaret Chamberlain, *Permissions Specialist (Pictures)*
Jeffrey Hermann, *Permissions Assistant*

Marco Di Vita, Graphix Group, *Typesetting*

> **Library of Congress Cataloging-in-Publication Data**
> Engelbert, Phillis.
>   The Complete Weather Resource / Phillis Engelbert.
>   p. cm.
>   Includes biographical references and index.
>   Contents: v. 1. Understanding weather — v. 2. Weather phenomena — v. 3. Forecasting & climate.
>   ISBN 0-8103-9788-9 (vol. 1 : alk. paper). — ISBN 0-8103-9789-7 (vol. 2 : alk. paper). — ISBN 0-8103-9790-0 (vol. 3 : alk. paper).
>   — ISBN 0-8103-9787-0 (set : alk. paper)
>   1. Weather. 2. Climatology. 3. Meteorology.   I. Title.
> QC981.E66   1997
> 551.5—dc21   97-6930
>   CIP

This publication is a creative work copyrighted by U•X•L and fully protected by all applicable copyright laws, as well as by misappropriation, trade secret, unfair competition, and other applicable laws. The authors and editors of this work have added value to the underlying factual material herein through one or more of the following: unique and original selection, coordination, expression, arrangement, and classification of the information. All rights to this publication will be vigorously defended.

Copyright © 1997
U•X•L An Imprint of Gale Research

All rights reserved, including the right of reproduction in whole or in part in any form.

 This book is printed on acid-free paper that meets the minimum requirements of American National Standard for Information Sciences—Permanence Paper for Printed Library Materials, ANSI Z39.48-1984.

Printed in the United States of America

10 9 8 7 6 5 4 3 2

TABLE OF CONTENTS

Reader's Guide VII
Words to Know IX
Picture Credits XLIII

VOLUME 1: UNDERSTANDING WEATHER

1: What Is Weather? 1
2: Clouds 75
3: Fog 109
4: Local Winds 121
Sources 151
Index 161

VOLUME 2: WEATHER PHENOMENA

5: Precipitation 181
6: Thunderstorms 211
7: Tornadoes 243
8: Hurricanes 262
9: Temperature Extremes, Floods, and Droughts 291
10: Optical Effects 317
Sources 339
Index 349

Table of Contents

VOLUME 3: FORECASTING AND CLIMATE

- 11: Forecasting 369
- 12: Climate 448
- 13: Human Activity and the Future 500
- Sources 521
- Index 531

Reader's Guide

The Complete Weather Resource presents under one title the most comprehensive survey of weather and weather-related topics to date, and it provides the clearest possible explanations for the many complex weather processes. The writing in *The Complete Weather Resource* is nontechnical and is geared to challenge, not overwhelm, students.

Volume 1, "Understanding Weather," focuses on basic atmospheric processes such as global and local winds, air pressure, heat and temperature, cloud formation, and front and storm formation.

"Weather Phenomena," Volume 2, offers in-depth discussions on precipitation, thunderstorms, tornadoes, hurricanes, temperature and precipitation extremes, and rainbows and other optical phenomena.

The third volume, "Forecasting and Climate," introduces the reader to several facets of these two topics. The "Forecasting" section provides information on state-of-the-art forecasting equipment and explains how to create forecasts using a backyard weather center. The "Climate" section describes climates of the world, changes in global climate throughout history, and reasons for climate change. Volume 3 concludes with a discussion on global warming, ozone depletion, and other environmental ills, as well as some steps that can be taken to protect the planet.

Scope and Format

The Complete Weather Resource is organized into chapters that are divided into topics and subtopics. The text is interspersed with boxes containing experiments, biographies, interesting weather facts, examples of extreme weather, and more. Approximately 140 photos, plus more

Reader's Guide

than 65 original illustrations and charts, keep the volumes lively and entertaining. Additionally, *The Complete Weather Resource* features cross-references, a glossary (glossary words are bolded throughout the text), and a cumulative index in all three volumes that provides easy access to the topics discussed throughout *The Complete Weather Resource.*

Advisors

Chris Gleason
Science Teacher, Greenhills School
Ann Arbor, Michigan

Ann Novak
Science Teacher, Greenhills School
Ann Arbor, Michigan

Dedication

The author dedicates this work to her brother, Jon Engelbert, for the sunshine he brings to her life.

Special Thanks

Special thanks go to Jane Hoehner, Senior Editor at U•X•L, for masterfully coordinating every aspect of this project; to David Newton for verifying the scientific soundness of this writing; to Pete Caplan, Meteorologist at the National Weather Service's National Center for Environmental Prediction, for answering numerous questions; and to Dr. Richard Wood, Meteorologist, for checking facts in the final manuscript.

Comments and Suggestions

We welcome your comments on this work as well as your suggestions for topics to be featured in future editions of *The Complete Weather Resource.* Please write: Editors, *The Complete Weather Resource,* U•X•L, 835 Penobscot Bldg., Detroit, Michigan 48226-4094; call toll-free: 1-800-877-4253; or fax: 313-877-6348.

WORDS TO KNOW

A

Absolute humidity: the amount of water vapor in the air, expressed as a ratio of the amount of water per unit of air.

Absolute zero: the temperature at which all motion ceases, -460°F (-273°C).

Accretion: the process by which a **hailstone** grows larger, by gradually accumulating cloud droplets as it travels through a cloud.

Acid fog: fog that is made more acidic by sulfuric and/or nitric acid in the air.

Acid rain: rain that is made more acidic by sulfuric and/or nitric acid in the air.

Adiabatic process: a process by which the temperature of a moving **air parcel** changes, even though no heat is exchanged between the air parcel and the surrounding air.

Advection fog: fog that forms when a warm, moist layer of air crosses over a cold surface.

Aerogenerator: an elevated windmill with very large blades.

Aerologist: someone who observes and gives reports of local atmospheric conditions.

Agricultural report: a specialized weather report, tailored to the needs of farmers, that includes current temperature, precipitation, and wind speed and direction, as well as frost warnings and predictions of temperature and precipitation for the days to come.

Words to Know

Air mass: a large quantity of air throughout which temperature and moisture content is fairly constant.

Air-mass thunderstorm: a relatively weak **thunderstorm** that forms within a single mass of warm, humid air.

Air-mass weather: unchanging weather conditions that result when a single **air mass** remains over a region for an extended period.

Air parcel: a small volume of air that has a consistent temperature and experiences minimal mixing with the surrounding air.

Air pollutant: any harmful substance that exists in the atmosphere at concentrations great enough to endanger the health of living organisms.

Air pressure: the pressure exerted by the weight of air over a given area of Earth's surface. Also called atmospheric pressure or barometric pressure.

Air stability: the temperature of the air at various heights, which determines whether an **air parcel** of a given temperature will rise, fall, or remain stationary.

Alberta Clipper: a dry, polar **air mass** that sweeps southward from Canada.

Altocumulus: clouds that looks like puffy masses and often appear in parallel rows or waves, that occupy an intermediate height in the **troposphere.**

Altostratus: nondescript, white, gray, or blue-gray, flat sheets of clouds that cover the entire sky and exist at an intermediate height in the **troposphere.**

Anabatic wind. *See* **Valley breeze**

Anemometer: an instrument used to measure wind speed, such as the **cup anemometer.**

Aneroid barometer: a type of **barometer** that consists of a vacuum-sealed metal capsule, within which a spring expands or contracts with changing **air pressure.**

Anticyclone: a weather system in which winds spiral clockwise, away from a high-pressure area.

Anvil: the flattened formation at the top of a mature **cumulonimbus** cloud.

Arctic climate: a **climate** type in which average temperatures remain below freezing, year-round, and the ground never thaws.

Arctic sea smoke: patchy, wispy **steam fog** that forms over unfrozen waters in arctic regions.

Atmospheric pressure. *See* **Air pressure**

Aurora: a bright, colorful display of light in the night sky, produced when charged particles from the sun enter Earth's atmosphere.

Avalanche: the cascading of some 100,000 tons of snow down a steep slope.

Aviation report: a specialized weather report, tailored to the needs of pilots, that provides information on the height of the clouds, visibility, and storm systems.

B

Backing wind: a wind that shifts direction, counterclockwise, with height.

Ball lightning: a mysterious form of **lightning** that is reported to look like a lighted sphere, ranging from .4 to 40 inches (1 to 100 centimeters) in diameter.

Banner cloud: a banner-shaped cloud that forms at a mountain's peak and drapes down over the **leeward slope.**

Barchan dune: a **sand dune** that, when viewed from above, resembles a crescent moon, with the tips of the crescent pointing downwind. Also called barchane dune, barkhan dune, or crescentic dune.

Barograph: an **aneroid barometer** that records changes in **air pressure** over time on a rotating drum.

Barometer: an instrument used to measure **air pressure.**

Barometric pressure. *See* **Air pressure**

Bead lightning: lightning that resembles a string of beads, that may be the result of the fragmentation of an **ionized channel.**

Blizzard: the most severe type of winter storm, characterized by winds of 35 mph (56 kph) or greater, large quantities of snow, and temperatures of 20°F (-6°C) or lower.

Blocking high. *See* **Blocking system**

Blocking low. *See* **Blocking system**

Words to Know

Blocking system: a whirling **air mass** containing either a high-pressure system (a blocking high) or a low-pressure system (a blocking low), that gets cut off from the main flow of **upper-air westerlies.**

Blowing snow: snow that has been lifted off the surface by the wind and blown about in the air.

Bolide: a large, rocky body from space, such as an asteroid or comet.

Bolide winter: a theoretical consequence of Earth being struck by a **bolide,** in which virtually all sunlight is blocked out by a thick dust cloud for a period of several months.

Buttes: steep, rocky hills of the American West.

C

Carbon dating: a technique, similar to **radioactive dating,** that uses an analysis of radioactive carbon to determine the age of rocks.

Carcinogens: cancer-causing agents.

Celsius scale: the temperature scale on which fresh water freezes at 0 degrees and boils at 100 degrees. To convert from Celsius to **Fahrenheit,** multiply degrees Celsius by 1.8, then add 32.

Cenozoic Era: the historical period from 65 million years ago to the present.

Chaos Theory: the theory that the weather, by its very nature, is unpredictable. Every time one atmospheric variable (such as heat, air pressure, or water) changes, every other variable also changes—but in ways that are out of proportion with the first variable's change.

Chinook: a dry, warm **katabatic wind** that blows down the eastern side of the Rocky Mountains, from New Mexico to Canada, in winter or early spring.

Chinook wall cloud: a solid bank of wispy, white clouds that appears over the eastern edge of the Rockies in advance of a **chinook.**

Cirriform: a wispy, feathery fair-weather cloud formation that exists at high levels of the **troposphere.**

Cirrocumulus: high, small, white, rounded, and puffy clouds that occur individually or in patterns resembling rippled waves, at high levels of the **troposphere.**

Words to Know

Cirrostratus: clouds that cover all or part of the sky, at high levels of the **troposphere,** in a sheet thin enough for the sun or moon to be clearly visible.

Cirrus: clouds at high levels of the **troposphere,** created by wind-blown ice crystals, that are so thin as to be nearly transparent.

Cirrus spissatus: tightly packed, icy **cirrus** cloud formations at the top of a **vertical cloud** that are dense enough to block out the sun.

Climate: the weather experienced by a given location, averaged over several decades.

Cloud-to-air lightning: lightning that travels between oppositely charged areas of a cloud and the surrounding air.

Cloud-to-cloud lightning: lightning that occurs within a single cloud or between two clouds.

Cloud-to-ground lightning: lightning that travels between a cloud and the ground.

Cloudburst: The heaviest type of **shower,** in which rain falls at a rate of 4 inches (10 centimeters) or more per hour.

Coalescence: the process by which an ice crystal grows larger. The ice crystal collides, and sticks together, with water droplets as the ice crystal travels down through a cloud.

Coastal flood: a **flood** that occurs along the coasts of a lake or ocean.

Cold cloud: a cloud within which ice crystals coexist with **supercooled water** droplets.

Cold fog. *See* **Freezing fog**

Cold front: the line behind which a cold **air mass** is advancing, and in front of which a warm air mass is retreating.

Cold occlusion: the most common type of **occluded front,** in which a cold **air mass** overtakes a warm air mass.

Compressional heating. *See* **Compressional warming**

Compressional warming: an **adiabatic process** by which an **air parcel** warms as it descends. The descending parcel is compressed by the increasing pressure of the surrounding air, which leads to a greater number of collisions between molecules. Also called compressional heating.

Condensation: the process by which water changes from a gas to a liquid.

Words to Know

Condensation nucleus: a tiny solid particle around which **condensation** of water vapor occurs.

Conduction: the transfer of heat from a fast-moving, warm molecule to a slow-moving, cold molecule.

Conservation of angular momentum: the principle that states that as the radius of a spinning object decreases, its speed increases, and vice versa.

Continental drift: the theory that over the last 200 to 250 million years, forces deep within Earth's core have caused a single huge continent to break apart and drift around the globe.

Contrails: **cirrus**-like markings in the sky, created by aircraft flying at 16,500 feet (5 kilometers) or higher. "Contrails" is an abbreviation for "condensation trails."

Convection: the upward motion of an **air mass** or **air parcel** that has been heated.

Convective cell: a unit within a **thunderstorm cloud** that contains **updraft**s and **downdraft**s.

Conventional radar: an instrument that detects the location, movement, and intensity of precipitation, and gives indications about the type of precipitation. It operates by emitting microwaves, which are reflected by precipitation. Also called radar.

Convergence: The movement of air inward, toward a central point.

Coriolis effect: the apparent curvature of large-scale winds, ocean currents, and anything else that moves freely across Earth, due to the rotation of Earth about its axis.

Corona: a circle of light centered on the moon or sun that is usually bounded by a colorful ring or set of rings.

Cosmic rays: invisible, high-energy particles that bombard Earth from space.

Crepuscular rays: bright beams of light that radiate from the sun and stretch across the sky.

Critical angle: the angle at which sunlight must strike the back of the raindrop, in order to be reflected back to the front of the drop.

Cumuliform: a puffy, heaped-up cloud formation.

Cumulonimbus: tall, dark, ominous-looking clouds that produce **thunderstorm**s. Also called thunderstorm clouds.

Words to Know

Cumulonimbus incus: a fully developed **cumulonimbus** cloud, the top of which extends to the top, or beyond the top, of the **troposphere.**

Cumulus: clouds that look like white or light-gray cotton puff-balls of various shapes.

Cumulus congestus: a tall **cumulus** cloud that is shaped like a head of cauliflower.

Cumulus humilis: the smallest species of **cumulus** cloud, which looks like small tufts of cotton.

Cumulus mediocris: a **cumulus** cloud of medium height with a lumpy top.

Cumulus stage: the initial stage of **thunderstorm** development, during which **cumulus** clouds undergo dramatic vertical growth. Also called developing stage.

Cup anemometer: an instrument used to measure wind speed. It consists of three or four cups positioned on their sides, connected by horizontal spokes to a cap that rotates freely on a pole.

Cyclogenesis: the process by which an **extratropical cyclone** is formed.

Cyclone: 1. a weather system in which winds spiral counterclockwise, in toward a low-pressure area. Also called storm. 2. the name for a **hurricane** that forms over the Indian Ocean.

D

Dart leaders: The series of dim **lightning** strokes that occur immediately after the original lightning stroke, that serve to discharge the remaining buildup of electrons near the base of the cloud.

Decay stage: the final stage of **tornado** development during which the tornado's funnel narrows, twists and turns, fragments, and dissipates.

De-icing: the process of spraying the wings of an aircraft with antifreeze before take-off, to prevent ice from accumulating on the wings.

Dendrite: a **sector plate** that has accumulated moisture and developed feathery branches on its arms. A dendrite is the most distinctive and most common type of **snowflake.**

Dendrochronology: the study of the annual growth of rings of trees.

Deposition: the process by which water changes directly from a gas to a solid, without first going through the liquid phase.

Words to Know

Deposition nuclei: tiny, solid particles suspended on clouds onto which water vapor molecules directly freeze, by the process of **deposition.**

Derecho: a destructive, straight-line wind, that travels faster than 58 mph (93 kph) and has a path of damage at least 280 miles (450 kilometers) long. Also called plow wind.

Desert climate: the world's driest **climate** type, with less than 10 inches (25 centimeters) of rainfall annually.

Desert pavement: hard, flat, dry ground and gravel that remains after all sand and dust has been eroded from a surface.

Developing stage. *See* **Cumulus stage**

Dew: clumps of water molecules that have condensed on a cold surface.

Dew point: the temperature at which a given parcel of air reaches its **saturation point** and can no longer hold water in the vapor state.

Diamond dust. *See* **Ice fog**

Diffraction: the slight bending of sunlight or moonlight around water droplets or other tiny particles.

Dispersion: the selective **refraction** of light that results in the separation of light into the spectrum of colors.

Dissipating stage: the final stage of a **thunderstorm,** during which the rain becomes light, the wind weakens, and the thunderstorm cloud begins to **evaporate.**

Divergence: the movement of air outward, away from a central point.

The Doctor: a special name given to the **sea breeze** in some tropical areas, because it brings relief from the oppressive heat.

Doldrums: the cloudy, rainy zone near the equator where the **trade winds** coming from north and south meet and nearly cancel each other out.

Doppler radar: a sophisticated type of radar that relies on the Doppler effect—the change in frequency of waves emitted from a moving source—to determine wind speed and direction, as well as the direction in which precipitation is moving.

Downburst: an extremely strong, localized **downdraft** beneath a **thunderstorm,** that spreads horizontally when it hits the ground, destroying objects in its path.

Downdraft: a downward blast of air from a **thunderstorm cloud**, felt at the surface as a cool gust.

Words to Know

Drifting snow: loose snow that has been swept into large piles, or "drifts," by strong winds.

Drizzle: precipitation formed by raindrops between .008 inches and .02 inches in diameter.

Dropwindsonde: a device, similar to a **radiosonde,** that is released by an aircraft and transmits atmospheric measurements to a radio receiver.

Drought: an extended period of abnormal dryness.

Dry adiabatic lapse rate: the constant rate at which the temperature of an unsaturated **air parcel** changes as it ascends or descends through the atmosphere. Specifically, air cools by 5.5°F for every 1,000 feet it ascends and warms by 5.5°F for every 1,000 feet it descends.

Dry-bulb thermometer. *See* **Thermometer**

Dry tongue: a layer of cold, dry, air that exists at an altitude of 10,000 feet (3,048 meters) and is necessary for the formation of a **supercell storm.**

Dust Bowl: the popular name for the approximately 150,000 square-mile-area (400,000-square-kilometer-area) in the South Great Plains region of the United States, characterized by low annual rainfall, a shallow layer of topsoil and high winds.

Dust devil: a spinning **vortex** of sand and dust that is usually harmless but may grow quite large. Also called a whirlwind.

Dust-whirl stage: the first stage in the formation of a **tornado,** marked by the emergence of a short **funnel cloud** and the swirling of debris on the ground.

E

Eccentricity: the alternating change in shape of Earth's orbit between a circle and an ellipse.

Eddies: small **air parcel**s that flow in a pattern that is different than the general air flow.

Ekman Spiral: the changing direction of the flow of ocean water along a vertical gradient.

El Niño: the annual, brief period during which the normally cold waters off the coast of Peru are made warmer by the arrival of warm waters from the equatorial region.

Words to Know

El Niño/Southern Oscillation (ENSO): a period during which a **major El Niño event** and a **Southern Oscillation** both occur. These two phenomena are connected because the warming of the waters off the coast of Peru lowers the **air pressure** in the eastern Pacific. As a result, the air pressure in the western Pacific rises.

Electromagnetic radiation: radiation that transmits energy through the interaction of electricity and magnetism.

Electromagnetic spectrum: the array of **electromagnetic radiation,** which includes radio waves, infrared radiation, visible light, ultraviolet radiation, X-rays, and gamma rays.

Enhanced greenhouse effect. *See* **Global warming**

Ensemble forecasting: A forecasting method takes into account the predictability of the behavior of atmosphere at the time a forecast is made.

Entrainment: the process by which cool, **unsaturated air** next to a **thunderstorm cloud** gets pulled into the cloud during the **mature stage** of a thunderstorm.

Equinoxes: the days marking the start of spring and fall and the two days of the year in which day and night are most similar in length.

Evaporation: the process by which water changes from a liquid to a gas.

Evaporation fog: fog that is formed when water vapor evaporates into cool air and brings the air to its **saturation point.**

Exosphere: the outermost layer of Earth's atmosphere, starting about 250 miles (400 kilometers) above ground, in which molecules of gas break down into atoms and, due to the lack of gravity, escape into space.

Expansional cooling: an **adiabatic process** by which an **air parcel** cools as it rises. The cooling is due to decreasing **air pressure** with altitude, which allows the air parcel to expand and leads to a smaller number of collisions between molecules.

Extratropical cyclone: a storm system that forms outside of the tropics and involves contrasting warm and cold **air mass**es.

Eye: the calm circle of low-pressure that exists at the center of a **hurricane.**

Eye wall: the region of a **hurricane** immediately surrounding the **eye,** and the strongest part of the storm. The eye wall is a loop of **thunderstorm cloud**s that produce torrential rains and forceful winds.

F

Fahrenheit scale: the temperature scale on which fresh water freezes at 32 degrees and boils at 212 degrees. To convert from Fahrenheit to **Celsius,** subtract 32 from degrees Fahrenheit, then divide by 1.8.

Fair-weather waterspout: a relatively harmless **waterspout** that forms over water and arises either in conjunction with, or independently of, a **severe thunderstorm.** Also called non-tornadic waterspout.

Fall streaks: ice crystals that fall from a cloud, **sublimate** into dry air, and never reach the ground.

Fata Morgana: a special type of **superior mirage** that takes the form of spectacular castles, buildings, or cliffs, rising above cold land or water.

Ferrel cell: an atmospheric cell through which **westerlies** and **upper-air westerlies** circulate. Air rises at 60 degrees latitude and sinks at 30 degrees latitude, North and South.

Fetch: the distance over water that the wind blows, which is used to calculate the height of waves.

Flash flood: a sudden, intense, localized flooding caused by persistent, heavy rainfall or the failure of a levee or dam.

Flood: the inundation of normally dry land with water.

Flurries: the lightest form of snowfall, which is brief and intermittent and results in little accumulation

Foehn: a warm, dry **katabatic wind** similar to the **chinook** that flows down from the Alps onto the plains of Austria and Germany.

Fog: a cloud that forms near or on the ground.

Fog stratus: a layer of **fog** that hovers a short distance above ground, without touching the ground. Also called **high fog.**

Forked lightning: lightning that results when a **return stroke** originates from two different places on the ground at once.

Freezing drizzle: drizzle comprised of **supercooled water** that freezes on contact with a cold surface.

Freezing fog: fog comprised of **supercooled water** droplets that freeze on contact with a cold surface. Also called cold fog.

Freezing nucleus: a tiny particle of ice or other solid onto which **supercooled water** droplets can freeze.

Words to Know

Freezing rain: rain comprised of **supercooled water** that freezes on contact with a cold surface.

Front: the dividing line between two **air mass**es.

Frontal fog: a type of **evaporation fog** that forms when a layer of warm air rises over a shallow layer of colder surface air. Also called precipitation fog.

Frontal system: a weather pattern that accompanies an advancing **front.**

Frontal thunderstorm: a **thunderstorm** that forms along the edge of a **front.**

Frontal uplift: the upward motion of a warm **air mass** caused by an advancing cold air mass.

Frost: ice that forms on a cold surface when the air directly above the surface reaches the **frost point.**

Frost point: the temperature at which a given **air parcel** reaches its **saturation point** and thus can no longer hold water in the vapor state, provided that the temperature is below freezing.

Frostbite: the freezing of the skin.

Fujita Intensity Scale: a scale that measures **tornado** intensity, based on wind speed and the damage created. Also called Fujita-Pearson Scale or Fujita Scale.

Funnel cloud: a cone-shaped **tornado** that hangs well below the base of a **thunderstorm cloud.**

G

Geostationary satellite: a **weather satellite** that remains "parked" above a given point on the equator, traveling at the same speed as Earth's rotation about 22,300 miles (35,900 kilometers) above the surface.

Glaze: a layer of clear, smooth ice on a cold surface formed when **supercooled water** strikes the surface and spreads out.

Global warming: the theory that average temperatures around the world have begun to rise, and will continue to rise, due to an increase of certain gases, called **greenhouse gases,** in the atmosphere. Also called enhanced greenhouse effect.

Global water budget: the balance of the volume of water coming and going between the oceans, atmosphere, and continental landmasses.

Words to Know

Glory: a set of colored rings that appears on the top surface of a cloud, directly beneath the observer. A glory is formed by the interaction of sunlight with tiny cloud droplets and is most often viewed from an airplane.

Graupel. *See* **Snow pellets**

Green flash: a very brief flash of green light that appears near the top edge of a rising or setting sun.

Greenhouse effect: the warming of Earth due to the presence of **greenhouse gases,** which trap upward-radiating heat and return it to Earth's surface.

Greenhouse gases: gases that trap heat in the atmosphere. The most abundant greenhouse gases are water vapor and carbon dioxide. Others include methane, nitrous oxide, and chlorofluorocarbons.

Ground blizzard: the drifting and blowing of snow that occurs after a snowfall has ended.

Ground fog: a very shallow layer of **radiation fog** that exists just above the ground.

Gust front: the dividing line between cold **downdraft**s and warm air at the surface, characterized by strong, cold, shifting winds.

Gyres: large, circular patterns of **ocean currents.**

H

Haboob: a tumbling black wall of sand that has been stirred up by cold **downdraft**s along the leading edge of a **thunderstorm** or **cold front,** that occurs in north-central Africa and, rarely, in the southwestern United States.

Hadley cell: an atmospheric cell through which **trade winds** circulate. Air rises at the equator and sinks at 30 degrees latitude, North and South.

Hail: precipitation comprised of **hailstone**s.

Hailstone: frozen **precipitation** that is either round or has a jagged surface, is either totally or partially transparent, and ranges in size from that of a pea to that of a softball.

Hair hygrometer: an instrument that measures **relative humidity.** It uses hairs (human or horse), which grow longer and shorter in response to changing humidity.

Words to Know

Halo: a thin ring of light that appears around the sun or the moon, caused by the **refraction** of light by ice crystals.

Haze: the uniform, milky white appearance of the sky that results when humidity is high and there are a large number of particles in the air.

Heat burst: a sudden, short-lived, dramatic warming of the air that is produced in the wake of a dissipating **thunderstorm.**

Heat cramps: muscle cramps or spasms, usually afflicting the abdomen or legs, caused by exercising in hot weather.

Heat equator: the warmest part of the equatorial zone, which lies most directly beneath the sun. Also called intertropical convergence zone (ITCZ).

Heat exhaustion: a form of mild shock that results when fluid and salt are lost through heavy perspiration.

Heat lightning: lightning from a storm that is too far away for its accompanying **thunder** to be heard.

Heat stroke: a life-threatening condition that sets in when **heat exhaustion** is left untreated and the body has exhausted its efforts to cool itself. Also called sunstroke.

Heat syncope: fainting that is the result of a rapid drop in blood pressure, that sometimes occurs while exercising in hot weather.

Heat wave: an extended period of high heat and humidity.

Heating-degree-days: the number of degrees difference between the day's mean temperature and an arbitrarily selected temperature at which most people set their thermostats. The number of heating-degree-days in a season is an indicator of how much heating fuel has been consumed.

Heavy rain: precipitation that falls at a rate greater than .3 inches (.76 centimeters) per hour.

Heavy snow: snowfall that reduces visibility to .31 miles (.5 kilometers) and yields, on average, 4 inches (10 centimeters) or more in a twelve-hour period or 6 inches (15 centimeters) or more in a twenty-four-hour period.

High fog. *See* **Fog stratus**

Highland climate. *See* **Mountain climate**

Hoar frost: frost that is formed by the process of **sublimation.** Also called true frost.

Hollow-column: a **snowflake** in the shape of a long, six-sided column.

Holocene epoch: the second part of the **Cenozoic Era,** from 10,000 years ago to the present.

Horse latitudes: a high-pressure belt that exists at around 30 degrees latitude, North and South, where air from the equatorial region descends and brings clear skies.

Humid subtropical climate: a **climate** type that has hot, muggy summers and mild, wet winters.

Humiture index: an index that combines temperature and **relative humidity** to determine how hot it actually feels and, consequently, how stressful outdoor activity will be. Also called temperature-humidity index or heat index.

Hurricane: the most intense form of **tropical cyclone.** A hurricane is a storm made up of a series of tightly coiled bands of **thunderstorm cloud**s, with a well-defined pattern of rotating winds and maximum sustained winds greater than 74 mph (119 kph).

Hydrologic cycle. *See* **Water cycle**

Hygrometer. *See* **Psychrometer**

Hypothermia: a condition characterized by a drop in core body temperature from the normal 98.6°F to below 95°F.

I

Ice age: a period during which significant portions of Earth's surface were covered with ice.

Ice fog: fog comprised of ice crystals that forms at temperatures below -22°F (-30°C). Also called diamond dust.

Ice pellets: Frozen raindrops formed by **precipitation** that first pass through a warm layer of air and melt, after which they enter a layer of freezing air and re-freeze.

Ice storm: a heavy downpour of **freezing rain** that deposits a layer of **glaze** more than an inch thick on solid objects it encounters.

Inferior mirage: a **mirage** that appears as an inverted, lowered image of a distant object. It forms in hot weather.

Instrument shelter: a ventilated wooden box on legs that is used to store and protect weather instruments outdoors. Also called a Stevenson screen or weather shack.

Words to Know

Insulator: a substance through which electricity does not readily flow.

Interglacial period: a relatively warm period that exists between two **ice ages**.

International weather symbols: the internationally accepted set of symbols used by meteorologists to describe many atmospheric conditions.

Intertropical convergence zone (ITCZ). *See* **Heat equator**

Inversion: an increase in air temperature with height.

Ionized: the condition of an object that has a positive or negative electrical charge.

Ionized channel: the path between a cloud and the ground, through which electrons flow, that is created when the **return stroke** of **lightning** contacts the **stepped leader.**

Ionosphere: the region of upper **mesosphere** and lower **thermosphere** in which molecules become **ionized** by X-rays and ultraviolet rays that exist in solar radiation.

Iridescence: an irregular patch of colored light on a cloud.

Isobar: an imaginary line that connects areas of equal **air pressure** that have undergone a **reduction to sea level.**

Isotherm: an imaginary line connecting areas of similar temperature.

J

Jet streams: narrow bands of fast winds that zip through the top of the **troposphere** in a west-to-east direction at speeds between 80 and 190 mph (128 and 305 kph).

K

Katabatic wind: a downhill wind that is considerably stronger than a **mountain breeze.**

Kelvin-Helmholtz clouds: thin clouds produced by **wind shear,** that look like a series of breaking ocean waves, in the upper levels of the **troposphere.**

Kinetic energy: the energy of motion.

L

Lake breeze: a **sea breeze**-type wind that can be felt on the edge of a large lake.

Lake-effect snow: a heavy snow that falls on the land downwind of the Great Lakes.

Land breeze: the gentle wind that blows from the shore to the water, due to differences in **air pressure** above each surface, at night.

Landfall: the passage of a **hurricane** from the ocean onto land.

Latent heat: the energy that is either absorbed by or released by a substance as it undergoes a phase change.

Latitude: an imaginary line encircling Earth, parallel to the equator, that tell one's position North or South on the globe.

Leeward slope: the eastward side of the mountain, on which cold air descends, producing dry conditions.

Lenticular cloud: a disc-shaped cloud that forms downwind of a mountain and remains in the sky for an extended period of time.

Lightning: a short-lived, bright flash of light during a **thunderstorm** that is produced by a 100-million-volt electrical discharge in the atmosphere.

Lightning rod: a metal pole that is attached to the tallest point of a building and connected, by an insulated conducting cable, to a metal rod buried deep the ground.

Local winds. *See* **Mesoscale winds.**

Longitude: an imaginary line encircling Earth, perpendicular to the equator, that tell one's position East or West on the globe.

M

Macroburst: a **downburst** that creates a path of destruction on the surface greater than 2.5 miles (4 kilometers) wide. The winds of a macroburst travel at around 130 mph (210 kph) and last up to thirty minutes.

Major El Niño event: a one- or two-year-long period during which the ocean waters off the coast of Peru remain warm. This prolonged warming, which has a variety of negative ecological consequences, occurs once every three to seven years.

Words to Know

Mammatus: round, pouch-like cloud formations that appear in clusters and hang from the underside of a larger cloud.

Marine climate: a coastal **climate** type that is characterized by cool summers, mild winters, and low clouds. **Fog** and **drizzle** are present for much of the year.

Marine forecast: a specialized weather forecast, of interest to coastal residents and mariners, that gives projections of the times of high and low tide, wave height, wind speed and direction, and visibility.

Mature stage: 1. the stage of **thunderstorm** development that begins when the first drops of rain reach the ground and is characterized by heavy rain, strong winds, **lightning,** and sometimes **hail** and **tornado**es. 2. the stage of tornado development during which the funnel reaches all the way to the ground and the tornado is at its most destructive.

Maunder minimum: the stretch of years from 1645 to 1715, during which **sunspot** activity was at a very low level.

Maximum and minimum thermometers: thermometers that record the highest and lowest temperatures during an observation period.

Mediterranean climate: a **climate** type characterized by dry summers and rainy, mild winters.

Melting zone: the atmospheric height at which the air becomes warm enough for falling snow to turn to rain.

Meltwater equivalent: the water content of snow.

Mercurial barometer: a type of **barometer** that relies on changes in a column of mercury to measure **air pressure.**

Mesocyclone: a region of rotating **updraft**s created by **wind shear** within a **supercell storm,** that may be a precursor to a **tornado.**

Mesoscale convective complex (MCC): a group of **thunderstorm**s that forms a nearly circular pattern over an area that is about a thousand times the size of an individual thunderstorm.

Mesoscale winds: winds that blow across areas of the surface ranging from a few miles to a hundred miles in width. Also known as **local winds** or **regional winds.**

Mesosphere: the middle layer of Earth's atmosphere, that exists between 40 and 50 miles (65 and 80 kilometers) above ground.

Mesozoic Era: the historical period from 225 million years ago to 65 million years ago, best known as the age of the dinosaurs.

Words to Know

Meteorology: the scientific study of the atmosphere and atmospheric processes, namely weather and climate.

Microburst: a very intense **downburst,** with winds that may exceed 167 mph (270 kph), that creates a path of destruction on the surface from several hundred yards wide to 2.5 miles (4 kilometers) wide.

Middle latitudes: the regions of the world that lie between the latitudes of 30 degrees and 60 degrees, North and South. Also called temperate regions.

Milankovitch theory: the theory stating that there are three types of variations in Earth's orbit that, taken together, can be linked with warm and cold periods throughout history. These variations include: the shape of Earth's orbit, the direction of tilt of its axis, and the degree of tilt of its axis.

Mirage: an optical illusion in which an object appears in a position that differs from its true position or in which a nonexistent object, such as a body of water, appears.

Mist: condensation that occurs in the low-lying air, in which visibility is greater than 1 kilometer.

Moist adiabatic lapse rate: the variable rate at which the temperature of a saturated **air parcel** changes as it ascends or descends through the atmosphere.

Monsoon climate: a **climate** type that is warm year-round with very rainy summers and relatively dry winters.

Moon dogs: patches of light, similar to **sundogs,** seen around a very bright, full moon.

Mountain breeze: a gentle downhill wind that forms at night as cold, dense, surface air travels down a mountainside and sinks into the valley. Also called gravity wind or drainage wind.

Mountain climate: the series of **climate** types that are found at various points ascending a mountainside. The range climate types results because temperature decreases with altitude. Also called highland climate.

Mountain-wave clouds: a class of clouds, including **lenticular cloud**s and **banner cloud**s, that are generated when moist wind crosses over a mountain range.

Multicell storm: a **thunderstorm** that contains several **convective cell**s.

Multi-vortex tornado: a **tornado** in which the **vortex** divides into several smaller vortices called **suction vortices.**

Words to Know

N

NEXRAD: Acronym for Next Generation Weather Radar, the network of 156 high-powered **Doppler radar** units which cover the continental United States, Alaska, Hawaii, Guam, and Korea.

Nimbostratus: dark gray, wet-looking layers of clouds that cover all or a large part of the sky, at low levels of the **troposphere.**

Nimbus: a dark, rain-producing cloud formation.

Nor'easter: a strong, northeasterly wind that brings cold air, often accompanied by heavy rain, snow, or sleet, to the coastal areas of New England and the mid-Atlantic states. Also called northeaster.

Northern Hemisphere: the half of Earth that lies north of the equator.

Numerical forecasting: the use of mathematical equations, performed by computers, to predict the weather.

O

Obliquity: the angle of the tilt of Earth's axis in relation to the plane of its orbit.

Occluded front: a **front** formed by the interaction of three **air mass**es: one cold, one cool, and one warm. The result is a multi-tiered air system, with cold air wedged on the bottom, cool air resting partially on top of the cold air, and warm air on the very top.

Ocean currents: the major routes through which ocean water is circulated around the globe.

Organized convection theory: the most widely accepted model of how a **hurricane** forms. It includes the formation of large **thunderstorm cloud**s, the transformation of a low-pressure area aloft into a high-pressure area (with a resultant low-pressure area at the surface), and the successive formation of spiraling bands of thunderstorms.

Organizing stage: the second stage of **tornado** formation, during which the **funnel cloud** extends part way to the ground and increases in strength.

Orographic lifting: the upward motion of warm air that occurs when a warm **air mass** travels up the side of a mountain.

Orographic thunderstorm: a type of **air-mass** thunderstorm that's initiated by the flow of warm air up a mountainside. Also called mountain thunderstorm.

Overshooting: the condition in which powerful **updraft**s rise above the **troposphere** and penetrate the **stratosphere.**

Ozone hole: the region above Antarctica in which the ozone layer virtually disappears at the end of each winter.

Ozone layer: the layer of Earth's atmosphere, between 25 and 40 miles (40 and 65 kilometers) above ground, that filters out the sun's harmful rays. It consists of ozone, which is a form of oxygen that has three atoms per molecule.

P

Paleoclimatologist: a scientist who studies **climate**s of the past.

Paleozoic Era: the historical period from 570 million years ago to 225 million years ago.

Particulate matter: air pollutants in the form of tiny solid or liquid particles, that creates the most visible type of air pollution.

Permafrost: a layer of subterranean soil that remains frozen year-round.

Photochemical smog: a hazy layer of surface ozone that sometimes appears brown. It is produced when pollutants that are released by car exhaust fumes react with strong sunlight.

Photovoltaic cell: an instrument that converts sunlight to electricity.

Pileus: smooth cloud formations that are found at the top of **cumulus congestus** or **cumulonimbus** clouds. Also called cap clouds.

Polar cell: an atmospheric cell that caps each pole, extending to 60 degrees latitude North and South. Relatively warm **westerlies** are carried poleward, and cold **polar easterlies** are carried equatorward, through each polar cell.

Polar climate: a **climate** type that covers the extreme northern and southern portions of Earth. It encompasses both **tundra climate** and **arctic climate.**

Polar easterlies: Cold, global winds that travel across the polar regions, from the northeast to the southwest in the **Northern Hemisphere** and from the southeast to the northwest in the **Southern Hemisphere.**

Words to Know

Polar fronts: The belts that encircle Earth at about 60 degrees latitude North and South, where the **westerlies** encounter **polar easterlies.**

Polar-orbiting satellite: a **weather satellite** that travels in a north-south path, crossing over both poles just 500 to 625 miles (800 to 1,000 kilometers) above Earth's surface.

Pre-frontal squall line: a **squall line** that forms some 100 to 200 miles (160 to 320 kilometers) ahead of a **cold front.**

Pre-Holocene epoch: the early part of the **Cenozoic Era,** from 65 million years ago to 10,000 years ago.

Precambrian Era: the historical period beginning with the formation of Earth, around 4.6 billion years ago, and ending 570 million years ago.

Precession of the equinoxes: the reversal of the **season**s every 13,000 years. This occurs because Earth spins about its axis like a top in slow motion and wobbles its way through one complete revolution every 26,000 years.

Precipitation: water particles that originate in the atmosphere (usually referring to water particles that form in clouds) and fall to the ground.

Precipitation fog. *See* **Frontal fog**

Pressure gradient: the rate at which **air pressure** decreases with horizontal distance.

Pressure gradient force (PGF): the force that causes winds to blow from an area of high pressure to an area of low pressure, that is proportional to the **pressure gradient.**

Prevailing winds: the winds blowing in the direction that's observed most often during a given time period.

Primary air pollutant: an **air pollutant** that is emitted directly into the air.

Psychrometer: an instrument used to measure **relative humidity.** It consists of a **dry-bulb thermometer** and a **wet-bulb thermometer.** Also called hygrometer.

R

Radar. *See* **Conventional radar.**

Radiation fog: fog that forms when a warm, moist layer of air exists at the surface and drier air lies above.

Radiational cooling: the loss of heat by the ground, to the atmosphere.

Radioactive dating: a technique used to determine the age of rocks that contain radioactive elements, which works on the principle that radioactive nuclei emit high-energy particles over time.

Radiosonde: an instrument package carried aloft on a small helium- or hydrogen-filled balloon. It measures temperature, **air pressure,** and **relative humidity** from the ground to a maximum height of 19 miles (30 kilometers).

Rain band: a tightly coiled band of **thunderstorm cloud**s that spirals around the **eye** of a **hurricane.**

Rain gauge: a container that catches rain and measures the amount of rainfall.

Rain-shadow effect: the uneven distribution of precipitation across a mountain, with most of the precipitation falling on the **windward slope** and very little falling on the **leeward slope.**

Rainbow: an arc of light, separated into its constituent colors, that stretches across the sky.

Rainforest climate: a **climate** type that is warm and rainy all year long.

Rawinsonde: a **radiosonde** that emits a signal so that its location can be tracked by radar on the ground. It measures changes in wind speed and wind direction with altitude.

Reduction to sea level: a process that standardizes **air pressure** readings with regard to altitude, making it possible to isolate differences in air pressure over horizontal distances.

Reflection: the process by which light both strikes a surface, and bounces off that surface, at the same angle.

Refraction: the bending of light as it is transmitted between two transparent media of different densities.

Regional winds. *See* **Mesoscale winds**

Relative humidity: A measure of humidity as a percentage of the total moisture a given volume of air, at a particular temperature, can hold.

Resolution: a measure of the precision of a weather forecast. The higher the resolution, the smaller the area for which the forecast is relevant.

Return stroke: lightning that surges up from the ground to meet the **stepped leader,** when the stepped leader is about 325 feet (40 meters) above the ground.

Words to Know

Ribbon lightning: **lightning** that appears to sway from the cloud. It is produced when the wind blows the **ionized channel** so that its position shifts between **return stroke**s.

Ridge: a northward crest in the wave-like flow of **upper-air westerlies,** within which exists a high-pressure area.

Rime: an icy coating that contains trapped air and therefore appears whitish, on a solid surface.

Riming: the process by which water droplets freeze to a snowflake, trapping air pockets.

River flood: the overflowing of the banks of a river or stream. It may be caused by excessive rain, the springtime melting of snow, blockage of water flow due to ice, or the failure of a dam or aqueduct.

Roll cloud: a cloud that looks like a giant, elongated cylinder lying on its side, that is rolling forward. It follows in the wake of a **gust front.**

Rossby waves: Long waves that are components of **upper-air westerlies.** At any given time, the entire hemisphere is encircled by just two to five Rossby waves.

S

Saffir-Simpson Hurricane Intensity Scale: the scale that ranks **hurricane**s according to their intensity, using the following criteria: **air pressure** at the **eye** of the storm; range of wind speeds; potential height of the **storm surge;** and the potential damage caused.

Saltation: the wind-driven migration of particles along the ground and through the air.

Sand dune: a mound of sand that is comprised of billions of sand grains, produced by a strong wind blowing in a fairly constant direction over time.

Sand ripples: wavy designs, running perpendicular to the direction of the wind, formed by the motion of sand along the surface of a **sand dune.**

Santa Ana winds: warm, dry, easterly or northeasterly winds that blow through southern California at a speed of at least 29 mph (46 kph).

Sastrugi: snow ripples (similar to **sand ripples**), up to 20 inches (50 centimeters) high, that form in Antarctica and other very cold places.

Words to Know

Saturated: air that has 100 percent **relative humidity.**

Saturation point: the point at which a given volume of air contains the maximum possible amount of water vapor. The addition of more water vapor at that point will result in **condensation.**

Savanna climate: a **climate** type that is warm year-round with rainy summers and drought-prone dry winters, and receives less yearly rainfall than a **monsoon climate.**

Scattering: multi-directional **reflection** of light by minute particles in the air.

Sea breeze: the gentle wind that blows from over the sea to the shore during the day, due to differences in **air pressure** above each surface.

Sea fog: a type of **advection fog** that only occurs at sea and in coastal areas. It is produced by the interaction of two adjacent **ocean currents** that have different temperatures.

Season: a period of year characterized by certain weather conditions, such as temperature and **precipitation,** as well as the number of hours of sunlight each day.

Secondary air pollutant: an **air pollutant** produced when a **primary air pollutant** undergoes chemical reactions with water, sunlight, or other pollutants.

Sector plate: a starry-shaped **snowflake.**

Seif dune: a very steep **barchan dune** with a very pronounced crescent shape, that either exists singly or in a connected line.

Semipermanent highs and lows: the four large pressure areas (two high-pressure and two low-pressure), situated throughout the **Northern Hemisphere,** that undergo slight shifts in position, and major changes in strength, throughout the year.

Severe blizzard: a **blizzard** in which wind speeds exceed 45 mph (72 kph), snowfall is heavy, and the temperature is no higher than 10°F (-12°C).

Severe thunderstorm: a **thunderstorm** that produces some combination of high winds, **hail, flash flood**s, and **tornado**es.

Sheet lightning: lightning that illuminates a cloud or a portion of a cloud.

Shelf cloud: a fan-shaped cloud with a flat base that forms along the edge of a **gust front.**

Words to Know

Short waves: The approximately twelve ripples that exist within each **Rossby wave.**

Shower: a spell of heavy, localized rainfall, that only occurs in warm weather.

Shrinking stage: the stage of **tornado** development during which the tornado's funnel narrows and tilts, and the tornado's path of destruction decreases in width.

Sinkhole: a dramatic example of **subsidence,** such as when roof of a cave collapses, that forms a large depression.

Sirocco: a hot dry, dusty southeasterly wind out of North Africa that travels across the Mediterranean Sea. It reaches Sicily and southern Italy as warm and humid wind.

Ski report: a specialized weather report that provides forecasts for popular ski destinations.

Sling psychrometer: an instrument that measures **relative humidity.** It consists of a **dry-bulb thermometer** and a **wet-bulb thermometer** mounted side by side on a metal strip, which rotates on a handle at one end.

Smog: a word created by combining "smoke" and "fog," that describes a thick layer of air pollution.

Snow dune: a large drift of snow, similar to a **sand dune.**

Snow fence: a device placed in fields and along highways that slows the wind and reduces the blowing and drifting of snow.

Snow grains: small, soft, white grains of ice that form within **stratus** clouds and only fall to the ground in small amounts. The frozen equivalent of **drizzle.**

Snow pellets: white pieces of icy matter that measure between .08 and .19 inches (.2 and .5 centimeters) in diameter. Snow pellets fall in showers and feel brittle and crunchy underfoot. Also called graupel or soft hail.

Snow ripples: long wavelike patterns in the snow that run perpendicular to the direction of the wind.

Snow squall: a brief but heavy snow shower, similar in intensity to a rain **shower,** accompanied by strong surface winds.

Snowflake: a hexagonal assemblage of ice crystals that is the basic unit of snow.

Snowroller: a lumpy, spherical or cylindrical mass of snow, generally less than 1 foot in diameter, formed by the wind.

Soft hail. *See* **Snow pellets**

Sounding: an analysis of temperature and humidity readings at various heights throughout the **troposphere.**

Southern Hemisphere: the half of Earth that lies south of the equator.

Southern Oscillation: a shifting pattern of **air pressure** between the eastern and western edges of the Pacific Ocean in the Southern Hemisphere.

Specific heat: the amount of heat required to raise one gram of a substance by 1°C (1.8°F).

Spontaneous nucleation: the process by which water freezes into ice crystals in the absence of **freezing nuclei,** which occurs at temperatures below -40°F (-40°C). Also called homogeneous nucleation.

Squall line: a band of thunderstorms that runs parallel to a **cold front,** either coinciding with the cold front or existing up to 200 miles (322 kilometers) in front of it.

Stable air layer: an atmospheric layer through which an **air parcel** cannot rise or descend.

Station circle. *See* **Weather station entry**

Stationary front: the dividing line between two stationary **air mass**es. It occurs when cold air comes in contact with warm air, yet neither side budges.

Steam devils: steam fog that forms in dense, rising, swirling columns over large bodies of water during the winter.

Steam fog: a type of **evaporation fog** that forms over a body of water.

Steppe climate: a semi-dry **climate** type that receives less than 20 inches (50 centimeters) of rainfall annually and exists in the **rain shadow** of a mountain range or at the edge of a desert.

Stepped leader: an invisible stream of electrons that initiates a **lightning** stroke. A stepped leader surges from the negatively charged region of a cloud, down through the base of the cloud, and travels in a stepwise fashion toward the ground.

Storm chaser: a professional or amateur weather-watcher who follows the path of a **tornado,** attempting to sight it and study its effects.

Words to Know

Storm surge: a wall of water, with huge waves, that sweeps on shore when the **eye** of the **hurricane** passes overhead.

Storm tide: the combined height of water that sweeps on shore in a **hurricane** due to the **storm surge** and the tide.

Stratiform: a cloud formation that appears as a continuous flat sheet or layer.

Stratocumulus: puffy clouds that exist in layers, at low levels of the **troposphere.**

Stratosphere: the second-lowest layer of Earth's atmosphere, from about 9 to 40 miles (15 to 65 kilometers) above ground.

Stratus: gloomy, gray, featureless sheets of clouds that cover the entire sky, at low levels of the **troposphere.**

Sublimation: the process by which water changes directly from a solid to a gas, without first going through the liquid phase.

Subpolar climate: a cold, northern type of **climate** that has long, harsh winters and short, cool summers.

Subsidence: the lowering of land in coastal areas, which makes them susceptible to flooding.

Suction vortices: small vortices within a single **tornado** that continually form and dissipate as the tornado moves along, creating the tornado's strongest surface winds.

Sundogs: one or two patches of light that appear on either or both sides of the sun. Sundogs are produced by the refraction of sunlight that shines through platelike ice crystals. Also called mock suns or parahelia.

Sunspot: a dark area of magnetic disturbance on the sun's surface.

Supercell storm: the most destructive and long-lasting form of **severe thunderstorm**, arising from a single, powerful **convective cell**. It is characterized by strong **tornado**es, heavy rain, and **hail** the size of golfballs or larger.

Supercooled water: water that remains in the liquid state below the freezing point.

Superior mirage: a cold-weather **mirage** that appears as a taller and closer, and sometimes inverted, image of a distant object.

T

Temperate climate: a **climate** type that has four distinct seasons with warm, humid summers and cold, snowy winters.

Terminal velocity: the constant speed at which an object falls when the upward force of air resistance equals the downward pull of gravity. An object can never fall at a rate faster than its terminal velocity. Also called maximum rate of fall.

Thermal: a pocket of rising, warm air that is produced by uneven heating of the ground.

Thermograph: an instrument consisting of a **thermometer** and a needle that etches on a rotating drum, continually recording the temperature.

Thermometer: an instrument used to measure temperature. It consists of a vacuum-sealed narrow glass tube with a bulb in the bottom containing mercury or red-dyed alcohol. Also called dry-bulb thermometer.

Thermosphere: the layer of Earth's atmosphere, between 50 and 200 miles (80 and 320 kilometers) above ground, in which temperatures reach 1,800°F (330°C).

Thunder: the sound wave that results when the intense heating due to **lightning** causes the air to expand explosively.

Thunderstorm: a relatively small but intense storm system, that produces moderate-to-strong winds, heavy rain, and **lightning,** and sometimes **hail** and **tornado**es.

Thunderstorm cloud. *See* **Cumulonimbus**

Topography: the shape and height of Earth's surface features.

Tornadic waterspout: a **tornado** that forms over land and travels over water. Tornadic waterspouts are relatively rare and are the most intense form of **waterspout**s.

Tornado: a rapidly spinning column of air that extends from a **thunderstorm cloud** to the ground. Also called a twister.

Tornado cyclone: a spinning column of air that protrudes through the base of a **thunderstorm cloud.**

Tornado family: a group of **tornado**es that develops from a single **thunderstorm.**

Tornado outbreak: the emergence of a **tornado family.** Tornado outbreaks are responsible for the greatest amount of tornado-related damage.

Words to Know

Trade winds: the winds that blow throughout the tropics, circulating air between the equator and 30 degrees latitude, North and South.

Transpiration: the process by which plants emit water through tiny pores in the underside of their leaves.

Transverse dunes: a series of connected **barchan dunes,** which appear as tall, elongated crescents of sand running perpendicular to the **prevailing wind.**

Traveler's report: a specialized weather report that tells what the weather is like at popular vacation spots and major cities around the world.

Tropical cyclone: a storm system that forms in the tropics, in the absence of **fronts.**

Tropical depression: the weakest form of **tropical cyclone,** characterized by rotating bands of clouds and **thunderstorm**s with maximum sustained winds of 38 mph (61 kph) or less.

Tropical disturbance: a cluster of **thunderstorm**s that is beginning to rotate.

Tropical squall cluster: a **squall line** that forms over tropical waters.

Tropical storm: a **tropical cyclone** weaker than a **hurricane,** with organized bands of rotating **thunderstorm**s and maximum sustained winds of 39 to 73 mph (63 to 117 kph).

Tropopause: the boundary between the **troposphere** and the **stratosphere,** between 30,000 and 40,000 feet (9 and 12 kilometers) above ground.

Troposphere: the lowest atmospheric layer, where clouds exist and virtually all weather occurs.

Trough: a southward dip in the wave-like flow of **upper-air westerlies,** within which exists a low-pressure area.

Tsunami: the largest type of water wave, generated by a submarine earthquake, landslide, or volcanic eruption.

Tundra climate: a **climate** type that has bitterly cold winters and cool summers. For at least one month of the year the average temperature is above freezing.

Typhoon: the name for a **hurricane** that occurs in the western North Pacific or China Sea region.

U

Unsaturated: air that has less than 100 percent **relative humidity.**

Unstable air layer: an atmospheric layer through which an **air parcel** can rise or descend.

Updraft: a column of air blowing upward, inside a **vertical cloud.**

Upper-air westerlies: global-scale, upper-air winds that flow in waves heading west-to-east (but also shifting north and south) through the **middle latitudes** of the **Northern Hemisphere.**

Upslope fog: fog formed by the slow passage of a moist **air parcel** up the side of a hill or mountain.

Upwelling: the rising up of cold waters from the depths of the ocean.

V

Valley breeze: an uphill wind that forms during the day as the valley air is heated and rises. Also called anabatic wind.

Valley fog: fog that forms when cold air sinks into a valley and, due to the presence of a river or stream, picks up moisture.

Vapor pressure: the pressure exerted by a vapor when it is in equilibrium with its liquid or solid. The vapor pressure determines the rate at which molecules of a substance will change phases.

Veering wind: a wind that shifts direction, clockwise, with height.

Ventifact: a rock, boulder, or canyon wall that has been sculpted by wind and wind-blown sand.

Vertical cloud: a cloud that develops upward to great heights. Vertical clouds are the products of sudden, forceful uplifts of small pockets of warm air.

Virga: Rain that falls from clouds but evaporates in mid-air under conditions of very low humidity.

Vortex: a vertical axis of extremely low pressure around which winds rotate.

Words to Know

W

Wall cloud: a roughly circular, rotating cloud that protrudes from the base of a **thunderstorm cloud** and is often a precursor to a **tornado.**

Warm clouds: clouds that exist in the tropics that are too warm to contain ice.

Warm front: the line behind which a warm **air mass** is advancing, and in front of which a cold air mass is retreating.

Warm occlusion: a rare type of **occluded front,** in which a relatively warm **air mass** overtakes a colder air mass.

Warning: a severe weather advisory that means that a storm has been sighted and may strike a specific area.

Watch: a severe weather advisory that means that while a storm does not yet exist, conditions are ripe for one to develop.

Water cycle: the continuous exchange of water between the atmosphere and the oceans and landmasses on the surface. Also called hydrologic cycle.

Waterspout: a rapidly rotating column of air that forms over a large body of water, extending from the base of a cloud to the surface of the water.

Weather aircraft: aircraft that carry weather instruments and collect data in the upper levels of the **troposphere.** They are primarily used to probe storm clouds, within which they measure temperature, **air pressure,** and wind speed and direction.

Weather forecast: a prediction of what the weather will be like in the future, based on present and past conditions.

Weather map: a map of a nation or group of nations, on which **weather station entries** are plotted. By looking at a weather map, a **meteorologist** can determine the locations of **fronts,** regions of high- and low-pressure, the dividing line between temperatures below freezing and above freezing, and the movement of storm systems. Also called surface analysis.

Weather modification: the use of artificial means to alter atmospheric phenomena.

Weather satellite: a satellite equipped with infrared and visible imaging equipment, that provides views of storms and continuously monitors weather conditions around the planet.

Words to Know

Weather station entry: the information collected at an individual weather station, recorded in **international weather symbols,** and placed on a **weather map.** Also called station circle.

Westerlies: Global-scale surface winds that travel from the southwest to the northeast in the **Northern Hemisphere,** and from the northwest to the southeast in the **Southern Hemisphere,** between about 30 and 60 degrees latitude.

Wet-bulb depression: the difference in temperatures measured by a **dry-bulb thermometer** and a **wet-bulb thermometer** at a given time.

Wet-bulb thermometer: a **thermometer** with wet muslin wrapped around the bulb, used to measure the temperature of **saturated** air.

Whirlwinds. *See* **Dust devils**

Whiteout: a condition in which falling, drifting, and blowing snow reduce visibility to almost zero.

Wind profiler: a specialized **Doppler radar,** resembling a giant metal checkerboard, that measures the speed and direction of **winds aloft.**

Wind shear: a condition in which a vertical layer of air is sandwiched between two other vertical layers, each of which is traveling at a different speed and/or direction, causing the sandwiched air layer to roll.

Wind sock: a cone-shaped cloth bag open on both ends, through which wind flows, that is used to determine the direction of the wind.

Wind turbine: a relatively small **windmill** with thin, propeller-like blades.

Wind vane: a free-swinging horizontal metal bar with a vertically oriented, flat metal sheet at one end and an arrow on the other end, that is used to determine the direction of the wind.

Wind waves: water waves that are driven by the wind.

Windchill equivalent temperature (WET): the temperature at which the body would lose an equivalent amount of heat, if there were no wind. Also called windchill index.

Windchill factor: the cooling effect on the body due to a combination of wind and temperature.

Winds aloft: winds that blow in the middle and upper levels of the **troposphere.**

Windward slope: the westward side of a mountain, on which warm air ascends, forms clouds, and yields precipitation.

PICTURE CREDITS

The photographs and illustrations appearing in *The Complete Weather Resource* were received from the following sources:

On the front cover of Volume 1: Virga (**FMA Research, Inc. Reproduced by permission.**); on the front cover of Volume 2: Aurora borealis (**JLM Visuals. Reproduced by permission.**); on the front cover of Volume 3: Power plant smokestack (**FMA Research, Inc. Reproduced by permission.**); on the back covers: weather map (**JLM Visuals. Reproduced by permission.**).

Courtesy of Gale Research: pp. 2, 12, 413, 501; **Corbis-Bettmann. Reproduced by permission:** pp. 8, 11, 16, 17, 205, 206, 379, 392, 475, 497, 519; **FMA Research, Inc. Reproduced by permission:** pp. 10, 50, 58, 59, 69, 78, 79, 80, 81, 82, 84, 85, 86, 87, 88, 89, 90, 93, 96, 97, 98, 99, 100, 102, 103, 105, 110, 111, 112, 113, 115, 116, 118, 119, 128, 130, 140, 147, 183, 185, 186, 193, 199, 201, 224, 233, 236, 255, 320, 332, 334, 335, 336, 337, 373, 383, 384, 396, 399, 402, 409, 411, 414, 416, 419, 432, 464, 476, 510, 520; **Mary Evans Picture Library. Reproduced by permission:** pp. 14, 15, 65; **Culver Pictures, Inc. Reproduced by permission:** pp. 20, 480; **Courtesy of Weather Service International Corporation. Reproduced by permission:** pp. 44, 282, 417, 418, 431; **JLM Visuals. Reproduced by permission:** pp. 52, 133, 305, 325, 380, 456, 459, 466, 468, 473, 484, 502, 507, 513; **Photograph by Cecil Keen. Reproduced by permission:** p. 101; **Photograph by Joanne Pease. Reproduced by permission:** p. 134; **Reuters/Corbis-Bettmann. Reproduced by permission:** pp. 142, 275, 311, 518; **Photograph by Mark Uliasz. Reproduced by permission:** p. 143; **Courtesy of Phillis Engelbert:** pp. 188, 462; **W. A. Bentley:** p.

Picture Credits

195; **UPI/Corbis-Bettmann. Reproduced by permission:** pp. 228, 294, 309, 312, 314, 492, 508, 517; **AP/Wide World Photos. Reproduced by permission:** p. 238; **National Severe Storms Laboratory. Reproduced by permission:** p. 244; **Photograph by Herbert Stein. Reproduced by permission:** p. 256; **National Aeronautics and Space Administration (NASA). Reproduced by permission:** p. 263; **Paramount/The Kobal Collection. Reproduced by permission:** p. 279; **Photograph by SIU. National Audubon Society Collection/Photo Researchers, Inc. Reproduced by permission:** p. 306; **Massachusetts Institute of Technology (MIT). Reproduced by permission:** pp. 374, 377; **National Weather Service. Reproduced by permission:** p. 428; **U. S. Department of Transportation. Reproduced by permission:** p. 429; **Courtesy of WNBC News Channel 4. Reproduced by permission:** p. 435; **Twentieth Century Fox/The Kobal Collection. Reproduced by permission:** p. 458.

All other diagrams were created by Accurate Art, Inc., Holbrook, New York.

PRECIPITATION

Precipitation is broadly defined as any form of liquid or solid water particles that originate in the atmosphere and reach Earth's surface. By this definition, precipitation includes **dew** and **frost,** as well as rain, snow, and ice (sleet and hail).

A narrower definition of precipitation is water particles that originate in the *clouds* and fall to the ground. By this definition, precipitation includes rain, snow, and ice. It is this definition that we will use in this chapter. Each of the three main forms of precipitation can be broken down into specific categories, according to the temperature of the air layers through which the precipitation passes, the size of the individual water particles, and the intensity with which it falls.

As we learned in the section on cloud formation in the chapter entitled "What Is Weather?" most precipitation (except in the tropics) originates in clouds as ice crystals. As an ice crystal descends through a cloud, it grows by collecting water vapor and **supercooled water** droplets. In the process, the ice crystal takes on the shape of a snowflake, a lump of snow or ice, or in severe **thunderstorm**s, hail. What happens next depends on the air temperature at various heights throughout the ice crystal's descent.

If the air temperature remains below freezing throughout the entire descent, the precipitation will reach the ground in the frozen state as snow. If the ice crystal passes through a layer of air above the freezing point of water, the ice crystal will melt and fall as rain. However, if the melting raindrop passes through a freezing layer, it will refreeze and reach the ground as **ice pellets.** Finally, if the ice crystal melts, only to

Precipitation

re-enter freezing air just at ground level, it will strike the ground as **freezing rain.**

Rain

Each raindrop is made up of a million or so microscopic cloud droplets. The average raindrop measures .04 to .24 inches (.10 to .61 centimeters) in diameter. Raindrops that grow larger than that become unstable and tend to break apart into smaller raindrops. According to a common misconception, "rain" is defined as liquid water that falls from the sky. In actuality, to be considered rain, a water drop must be larger than .02 inches (.05 centimeters). Precipitation consisting of anything smaller than that is called **drizzle.**

The main sources of rain are thick clouds with low bases, namely **nimbostratus** and **cumulonimbus** clouds. Occasionally rain falls from a thick layer of **altostratus** or tall **cumulus** clouds (see "Clouds," pages 81 and 88).

Some raindrops form in **warm cloud**s, clouds that are too warm for ice crystals to form. Although raindrops also collide with water droplets as they descend through a cloud, the collisions are less likely to result in **coalescence.** Hence, water drops do not grow as large as their ice-crystal counterparts. Raindrops from warm clouds are usually less than .08 inches (.2 centimeters) in diameter.

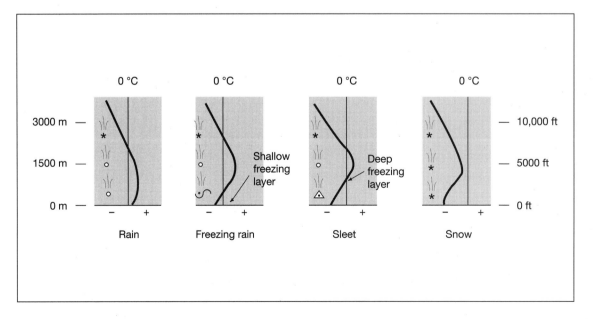

Figure 18: Vertical air temperature profiles through which various forms of precipitation descend.

The exception to this rule are raindrops that form in towering cumulus or cumulonimbus clouds in the tropics. In those clouds, where strong **updraft**s exists, a water drop may be blown from the bottom to the top of the cloud several times, growing larger each time, until it finally falls to the ground. Raindrops up to .32 inches (.8 centimeters) in diameter have been found falling from such clouds in Hawaii.

You may have noticed that visibility in the air improves following a rainfall. The reason for this is that precipitation has a cleansing effect on the air. As we learned in the chapter "What Is Weather?" water droplets in the air form around **condensation nuclei,** tiny particles of dust and debris. When precipitation falls, it removes these particles from the air.

Drizzle

As mentioned before, drizzle is precipitation made up of drops that are between .008 inches (.02 centimeters) and .02 inches (.05 centimeters) in diameter. These drops are just barely large enough to overcome the upward force of air resistance. Once they do, they slowly drift downward, sometimes taking over an hour to travel from the cloud to the ground.

There are two ways in which drizzle is produced. The first is that it falls from **stratus** clouds, which are low and exist in shallow layers less than 1.5 miles (about 2.5 kilometers) thick. In stratus clouds, water drops have far less opportunity to grow by coalescence than they do in rain-

Downpours from approaching thunderstorms in Colorado.

Precipitation

WEATHER REPORT: THE SHAPE OF A RAINDROP

Contrary to popular belief, raindrops are not tear-shaped or pear-shaped. As you can see in the diagram, raindrops vary in shape between a sphere and a lump, depending on their size.

Small raindrops, less than .08 inches (.2 centimeters) in diameter, are nearly spherical. Raindrops with diameters greater than .08 inches look more like oblong blobs, flattened at the bottom and rounded on the top, and are wider than they are tall. Large raindrops have been compared in appearance to a falling parachute, a mushroom cap, and a hamburger bun.

The shape of a raindrop is caused by surface tension. Surface tension is the attraction between water molecules at the surface. Surface tension forces molecules into the configuration with the smallest surface area, which is a sphere. In larger raindrops, this shape is distorted by the effect of **air pressure.** Air pressure is felt most strongly on the bottom of the drop and most weakly on the sides of the drop. It pushes the bottom upward, flattening it, while allowing the sides to expand.

If a raindrop becomes larger than .25 inches (.6 centimeters) in diameter, it will flatten out even more. At the same time it will become pinched on the top, into a bow-tie shape. The two halves of the bow tie will bulge and become more pronounced until the drop divides into two smaller, spherical drops.

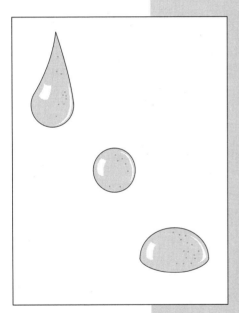

producing clouds. The second way in which drizzle forms is that it begins as rain, descends through dry air, and partially evaporates. By the time the raindrops reach the ground they have been reduced to the size of drizzle.

Drizzle may fall constantly, even for an entire day. The heaviest drizzle is produced where warm, moist air rises along the side of a mountain and forms **mountain wave clouds.** Drizzle that falls from these clouds can produce .4 inches (1 centimeter) of water per day.

The drops in drizzle fall very close together, which reduces visibility. In a heavy drizzle, it may be possible to see only 5/16 of a mile (.5 kilometers) ahead.

VIRGA

When humidity is very low, rain or snow may completely evaporate into the air during its descent. This creates streaks of falling water, called **virga.** Virga looks like dark fringes extending from the base of a cloud. It is not considered true precipitation since it doesn't reach the ground.

Virga often sets the stage for precipitation. It does this by increasing the humidity of the air into which it evaporates. Thus, water or ice that falls may subsequently make it to the ground.

SHOWERS

A **shower** is a spell of heavy, localized rainfall, that occurs only in warm weather. Showers fall from towering **cumuliform** clouds, which are produced by strong **convection** currents. Convection is the rising of pockets of warm air that occurs when Earth's surface is heated.

A shower occurs only while the shower-producing cumuliform cloud is overhead. It can last anywhere from two minutes to a half-hour,

Orographic stratus, a type of mountain wave cloud: thick stratus clouds form over the slopes of Maui in the Hawaiian Islands as moist, tropical trade wind air is forced upwards.

Precipitation

depending on wind speed and the size of the cloud. In an area where a series of cumuliform clouds exist, several showers may occur, separated by dry, even sunny, periods.

The area being showered at any given time is generally no larger than 4 to 5 square miles (10 to 13 square kilometers). Rain, in contrast, can fall on an area larger than 100 square miles (260 square kilometers) at a time and can last all day.

The reason that towering cumuliform clouds give rise to showers is that they are able to generate large quantities of raindrops quickly. The process that takes place is as follows: Ice crystals or water drops bounce between the top and bottom of the cloud numerous times, growing larger by coalescence on each trip. In most cases, by the time an ice crystal reaches the bottom of the cloud, the air is warm enough to cause the ice to melt.

When a water drop becomes too large, it breaks apart, forming smaller drops. These drops, in turn, get blown to the top of the cloud and the process of coalescence is repeated. A chain reaction ensues, producing more and more drops. This reaction continues until the **updraft**s weaken or change direction. Then a sudden, heavy shower falls.

The heaviest showers are called **cloudburst**s. To qualify as a cloudburst, precipitation must fall at a rate of 4 inches (10 centimeters) or more per hour. The heaviest showers occur in the tropics, where the air is

Virga, or falling rain that evaporates in the air.

Precipitation

A Key Reference To: Measuring the Intensity of Rainfall

The intensity of precipitation is the amount of water that falls over a given period of time. For instance, a cloudburst is the most intense form of rainfall because it releases a large amount of water to the ground in a very short period of time. On the other hand, a steady rain may produce the same amount of water as a cloudburst, but since it falls over a longer period, it has a lower intensity. (For methods of measuring precipitation, see "Forecasting," page 398.)

The following are commonly accepted definitions for categories of rainfall, based on intensity:
- <u>Heavy rain</u>: greater than .3 inches (.75 centimeters) of water falls per hour. Heavy rain appears to fall in sheets and greatly reduces visibility.
- <u>Moderate rain</u>: between .11 and .3 inches (.25 and .76 centimeters) of water falls per hour. While the rain does not fall in sheets, it still falls too fast to see individual raindrops.
- <u>Light rain</u>: up to .1 inches (.25 centimeters) of water falls per hour. Individual raindrops can be seen.
- <u>Trace</u>: rainfall is too light to measure.

The intensity of drizzle, which produces very small quantities of water, is measured in terms of visibility.
- <u>Heavy drizzle</u>: visibility is ≤ ¼ mile (.5 kilometers).
- <u>Moderate drizzle</u>: visibility is > ¼ mile but ≤ ½ mile (.5 and 1 kilometer).
- <u>Light drizzle</u>: visibility is > ½ mile (1 kilometer).

warm and moist and powerful convection currents produce huge **thunderstorm cloud**s.

Freezing Rain

Freezing rain, as its name suggests, is rain that freezes on the ground. Freezing rain begins its journey to the surface as snow. As the

Precipitation

snow descends, it encounters a layer of warm air and melts. Then, just above the ground, it travels through a shallow layer of subfreezing air. The raindrops don't have time to refreeze, but remain in the liquid state at temperatures below freezing. In other words, they become supercooled.

When the supercooled liquid makes contact with a cold surface, it spreads out and then freezes. It forms a layer of clear, smooth ice called **glaze.** If drizzle becomes supercooled and then freezes on the ground, it is known as **freezing drizzle.**

Freezing rain most often occurs in the winter, following a cold night in which the ground rapidly loses heat to the atmosphere through **radiational cooling.** As a result, an **inversion** is produced. An inversion exists when a layer of cold air is found next to the ground and a warmer layer of air lies above. Freezing rain creates a beautiful, but hazardous, coating of ice on trees, power lines, and roads.

A heavy downpour of freezing rain is called an **ice storm.** While the layer of glaze deposited by freezing rain is usually less than an inch thick, in ice storms it can be much thicker. For instance, an ice storm in northern Idaho in January 1961 produced an 8-inch-thick (20-centimeter-thick) layer of glaze. It has been estimated that during a severe ice storm, a 50-foot-tall (15-meter-tall) evergreen tree with an average width of 20 feet (6 meters) may be loaded down with five tons of ice.

A layer of glaze coats trees and shrubs following an ice storm.

Precipitation

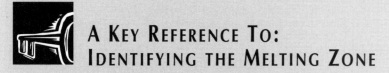

A KEY REFERENCE TO: IDENTIFYING THE MELTING ZONE

When rain is falling on a cold day, it is sometimes possible to see the height at which the precipitation changes from its original state—snow—to rain. To try this, look at the clouds that are in front of the sun, when the sun is near the horizon. You may see a dark upper layer of clouds, and underneath that, a light region. The transition area between dark and light regions is the **melting zone,** the height in the sky where the air warms and snow turns to rain.

The reason why the colder, upper level of air is darker than the warmer, lower level is that snow scatters sunlight better than does rain. Thus, less sunlight from the upper layer reaches your eye. The warmer, lower, rainy level appears lighter, in contrast.

In an ice storm, it is common for tree limbs and power and telephone lines to be knocked down. The ice is very difficult to drive on and causes traffic accidents. Over 85 percent of the deaths that occur in ice storms are traffic-related.

One of the most severe ice storms on record is that which struck a huge area, from Mississippi to New England, in January 1983. That storm resulted in an estimated 25 deaths and cut off power to more than 250,000 people.

Freezing rain is common in hilly or mountainous areas, where cold air sinks into the valleys. In the Appalachian mountains of Pennsylvania and West Virginia, freezing rain can fall on localized areas for long periods of time. The ice deposited by freezing rain usually melts within a few hours, although occasionally it can persist for days. The record for the longest-lasting glaze was set in 1969. In that year, ice remained on the trees for six weeks in Connecticut.

SNOW

Snow is precipitation that is common during the winter in **middle latitudes** and year-round on mountaintops. Its basic unit is the

Precipitation

WEATHER REPORT: RAINMAKING

People throughout the ages have sought ways to bring rain to moisture-deprived areas. In the 1800s and early 1900s people tried, unsuccessfully, to produce rain by ringing church bells or firing cannons into the air. In the mid-1940s, scientists began testing an experimental method called cloud seeding. Cloud seeding involves injecting particles into a cloud, which act as **freezing nuclei.** Cloud droplets adhere to the injected particles and fall to the ground as **precipitation.**

A requirement of cloud seeding is that clouds are already present. And those clouds must be tall enough so that their upper portions extend into regions where temperatures are below freezing. In other words, they must be **cold clouds.** That is because the droplets that stick to the injected particles must be **supercooled,** meaning they exist in the liquid state at temperatures below freezing. And supercooled water droplets are only found in cold clouds.

The earliest cloud seeding experiments, performed in 1946 by atmospheric scientist Vincent Schaefer, involved dropping crushed dry ice (carbon dioxide) pellets into the top of a cloud, from an airplane. Since dry ice is extremely cold (-108°F or -78°C), it cools the air around it, which produces more **condensation.** The dry ice pellets also act as freezing nuclei. In one experiment, Schaefer dropped three pounds of ground dry ice into an **altocumulus** cloud. Five minutes later, snow began falling from that cloud. The following year one of Schaefer's colleagues, Bernard Vonnegut, discovered that silver iodide makes a better cloud seeding agent than dry ice.

The 1950s saw a boom of rainmaking operations in **drought**-stricken areas around the world. These operations, however, failed to produce a significant increase in precipitation. Furthermore, they provoked the objections of many people who were concerned with the broader implications of manipulating the atmosphere. Would cloud seeding irreversibly alter the Earth's **water cycle**? Would it lead to uncontrollable **flood**s or prolonged droughts?

Despite these concerns, at the end of the 1950s a U.S. federal research program was launched on cloud seeding and other forms of

weather modification (the use of artificial means to alter atmospheric phenomena). Government agencies and private companies conducted many experiments, with inconclusive results.

Cloud seeding was the center of controversy more than once in the 1970s. In 1972, a spate of cloud seeding was followed by a **flash flood** in Rapid City, South Dakota, in which over 200 people lost their lives. While the cloud seeding and the flash flood may or may not have been linked, the tragedy was enough to deter people from further rainmaking experiments there. Cloud seeding was also a questionable practice used by the U.S. military during the Vietnam War. After the flooding of the Ho Chi Minh Trail, which supposedly came about as a result of cloud seeding, the practice was called into question in the U.S. Senate.

Does cloud seeding really work? Fifty years after the first cloud-seeding experiment, this question is still hotly debated. Some studies suggest that, under particular circumstances, cloud seeding can increase precipitation by 5 to 20 percent.

Some of the most impressive results have been obtained by seeding **cumulus** clouds. As droplets freeze to the injected particles, they release **latent heat,** which fuels the upward expansion of the clouds. And when a cumulus cloud develops vertically, it lasts longer and is more likely to produce precipitation. Another way in which cloud seeding seems to be effective is by seeding winter clouds that are already producing precipitation. This practice has been shown to increase the amount of snow falling from those clouds.

On the flip side, by overseeding a cloud, it's possible to reduce precipitation. When too many freezing nuclei are present, they remain too small to fall to the ground—even after all available supercooled droplets have frozen onto the freezing nuclei. The moisture in the cloud will then evaporate. For this reason, overseeding is used at airports to dissipate thick **fog.**

The practice of cloud seeding continues to the present day. Current expectations of what cloud seeding can produce, however, are quite humble compared to the expectations of researchers in the 1940s and 1950s. In the United States, cloud seeding is performed mainly in California. There, the goal of cloud seeding is to slightly increase the amount of rainfall or snowfall produced by storms.

Precipitation

WEATHER REPORT: ICE AND AIRCRAFT

Freezing rain and **supercooled water** droplets within clouds pose major hazards to aircraft. As an aircraft travels through clouds where the temperature is between 10 and 32°F (-12 and 0°C), it encounters a mixture of ice crystals and supercooled droplets. Where supercooled raindrops or large droplets in clouds come in contact with an aircraft, they spread out and form an even coating of ice, called **glaze.** Where the small droplets strike an aircraft, they freeze immediately without spreading, trapping air bubbles. The second type of icy coating, which appears white and weighs less than glaze, is called **rime.**

A layer of glaze, and to a lesser extent a layer of rime, affects an aircraft in several ways. First, it makes the plane heavy—sometimes so heavy that it is literally pulled from the sky. Second, if ice forms on the plane's wing or fuselage, it provides wind resistance and alters the plane's aerodynamics. If the engine's air intake opening is iced over, it can lead to a loss of power. Ice can also cause the failure of brakes, landing gear, or instruments.

Ice poses the greatest danger to small, single-engine or twin-engine planes. Ice affects jet airliners to a lesser extent since those planes spend most of their time above the clouds. Nonetheless, airline pilots are given warnings of, and strive to avoid, clouds that may contain supercooled droplets. As an added precaution, the wings of an aircraft are usually **de-iced,** meaning they are sprayed with a type of antifreeze, before take-off.

---

snowflake, which consists of many ice crystals joined to each other. Like raindrops, snowflakes come in various shapes and sizes. And, like rain, snow is categorized by the intensity with which it falls: from the lightest form, **flurries,** to the heaviest form, a **blizzard.**

In order to qualify as "snow," precipitation must remain frozen when it reaches the ground. This does not mean, however, that it can snow only when the surface air temperature is at or below freezing.

Precipitation

Rather, snow can remain frozen at temperatures above freezing, for distances up to 1,000 feet (300 meters), without melting. The exact air temperature at which snow turns to rain, depends on the humidity of the air.

At the beginning of a snowfall, when temperatures are above freezing, snow may turn to rain as it falls. If air is dry, rain rapidly evaporates into it. The process of **evaporation** draws heat from the air, leaving the air cooler than before the evaporation began.

As snow continues to fall from the cloud, it encounters lower temperatures, although temperatures may still be above freezing. If the snow enters above-freezing temperatures, it begins to melt. Water from the edge of a snowflake rapidly evaporates into the air, further cooling the air, as well as cooling the snowflake.

If cooled below the freezing point by the evaporative process, the snowflake will reach the ground intact. If not, it will melt and some of the water will evaporate, further lowering the air temperature. As long as air remains **unsaturated** (has less than 100 percent **relative humidity**), the cooling process will continue. However, once air becomes **saturated,** no net evaporation will occur, and cooling will cease.

What is the warmest surface air temperature in which precipitation can reach the ground as snow? A **wet-bulb thermometer** reading will tell you. The wet-bulb reading is the temperature of the air at its saturation point. As we discussed in the chapter on "Forecasting," one way to

Snow mounds in a snowbelt downwind of the Great Lakes.

Precipitation

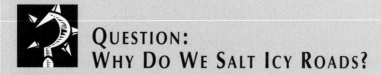
QUESTION: WHY DO WE SALT ICY ROADS?

Salt is applied to snowy, icy roads because it melts ice and prevents the water from re-freezing. Compared to other materials that are used to combat slippery roadways, such as sand and cinders, salt is relatively cheap and easy to apply. For these reasons, salt has been the road de-icing agent of choice since the 1960s.

The chemical composition of salt used on roads is sodium chloride (NaCl). When it comes in contact with water molecules, NaCl breaks down into one positively charged sodium ion (NA^{+2}) and two negatively charged chloride ions (Cl^-). A water molecule consists of one oxygen atom (O^{-2}) and two hydrogen atoms (H^+). The positively charged hydrogen atoms, are drawn to the negatively charged chloride atoms. And negatively charged oxygen atoms are drawn to the sodium atoms.

Sodium chloride thus causes the components of individual water molecules to disassociate. It bonds with the hydrogen and oxygen atoms so that hydrogen and oxygen are not free to recombine into water. The sodium and chloride ions also draw water molecules away from one another. Salt both prevents liquid water molecules from forming ice crystals and breaks up existing ice crystals.

Sodium chloride lowers the freezing point of water from 32°F (0°C) to 20°F (-6.7°C). At temperatures below 20°F, salt is no longer effective at melting ice.

Whereas salt provides an efficient means of melting snow and ice on roads, and has greatly contributed to highway safety, it also has a downside. Salt is bad for the environment. It kills vegetation along the side of the road and can seep down into wells, making the water undrinkable. Salt also causes vehicles to rust and bridges to corrode. For these reasons, salt is only applied in the minimum quantities necessary to get the job done.

determine relative humidity is to take the difference between the readings of dry-bulb and wet-bulb thermometers and then enter that figure on a chart (see "Forecasting," page 390).

As long as the wet-bulb temperature of the air is at freezing or below, falling snow is able to cool the air around it to the freezing point and can reach the ground intact. The wet-bulb reading can still be at the freezing point when the air temperature is at a maximum of 50°F (10°C). This means that, theoretically, snow can fall on the ground when the air is as warm as 50°F. This situation would be a very unusual, however, since the relative humidity would have to be only 11 percent. Such dry conditions are most often associated with much colder temperatures.

Snow has even been recorded falling at temperatures above 50°F when snowflakes are blown downward with a cold, dry **downdraft** from a towering **cumulonimbus** cloud.

SNOWFLAKES

The first person to study the intricate detail and uniqueness of snowflakes extensively was an American farmer named William Bentley (1865–1931). Beginning in 1880 when he was just fifteen years old, Bentley placed snowflakes under a microscope and photographed them. He continued this work for fifty years, making thousands of photographs. In 1931, he published a book with W. J. Humphreys, entitled *Snow Crystals,* that contained over 2,300 photos of snow and **frost.**

One thing that all snowflakes have in common is a hexagonal (six-sided) configuration. The structure of the ice crystals that make up a snowflake is also hexagonal. In fact, this basic shape can be traced back to water molecules. Because of the electrical attraction between water molecules, they take a hexagonal shape when they freeze.

A snowflake begins its existence as an ice crystal within a **cold cloud.** As it bounces between the bottom and top of the cloud, it grows by **coalescence** with supercooled water drops or by **deposition,** the freezing of water vapor molecules directly onto the ice crystal. As the ice crystal grows, it bonds with other ice crystals and assumes the shape of a snowflake (also called a "snow crystal"). When the snowflake becomes heavy enough, it descends to the surface.

Snowflakes come in a variety of shapes and sizes.

Precipitation

Snowflakes can exist in flat, plate-like forms; long, six-sided columns; or needles that are two hundred times longer than they are wide. They may also form starry shapes, called **sector plates.** When a sector plate accumulates moisture it may develop feathery branches on its arms. In this way, the most distinctive and most common snowflake, the **dendrite,** is formed. As dendrites travel through the cloud, they may combine with other dendrites, forming a wide array of complex patterns.

As illustrated in Figure 19, the shape of a snowflake depends upon the air temperature within the cloud where it is formed. Below -8°F (-22°C) snowflakes are **hollow columns**; between -8 and 3°F (-22 and -16°C) they are sector plates; between 3°F and 10°F (-16 and -12°C) they are dendrites; between 10°F and 14°F (-12 and -10°C) they are sector plates; between 14°F and 21°F (-10 and -6°C) they are hollow columns; between 21°F and 25°F (-6 and -4°C) they are needles; and over 25°F (-4°C) they are thin hexagonal plates.

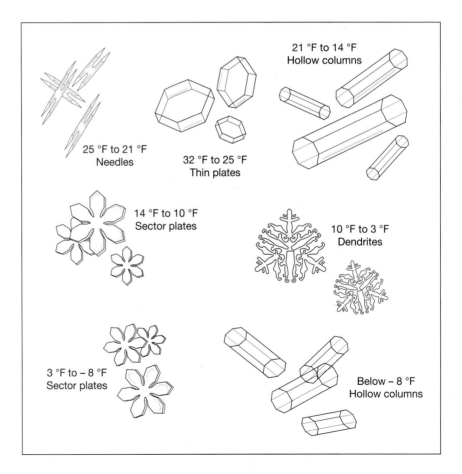

Figure 19: The shapes of snowflakes at various temperatures.

Precipitation

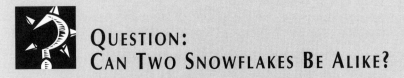
QUESTION: CAN TWO SNOWFLAKES BE ALIKE?

In order to answer this question, we must define "alike." "Alike" can mean that two snowflakes are the same, molecule for molecule, throughout. By this definition, the answer is no, two snowflakes cannot be alike. First of all, a snowflake contains over 180 billion water molecules. These molecules come together under many different conditions, making it all but impossible for any two snowflakes to have an identical configuration. In addition, water molecules are constantly freezing to and evaporating from snowflakes, meaning that snowflakes are constantly changing at the molecular level.

An alternate definition of "alike" is "identical in appearance." By this definition, the answer is yes, two snowflakes can be alike. This fact was discovered in 1989 by Nancy Knight, a cloud physicist with the National Center for Atmospheric Research. Knight collected snow samples while on board a research aircraft at a height of 20,000 feet (6 kilometers) over Wasau, Wisconsin. She discovered two hollow-column snowflakes, both 250 microns long and 170 microns wide (a micron is one-millionth of a meter—by way of comparison, a human hair is about 100 microns in diameter), proving that snowflakes can be identical in appearance. To this day, however, there is no record of identical dendrites.

You may wonder why dendrites are the most common form of snowflake, given that they only form within a narrow range of temperatures (10 and 14°F or -12 and -10°C). The reason is that dendrites form more rapidly than do other snowflake types.

In the dendrite-forming temperature range, the difference in **vapor pressure** between water droplets and ice crystals is greatest. Vapor pressure is the pressure exerted by a vapor when it is in equilibrium with its liquid or solid. Vapor pressure is greater over the surface of a water droplet than it is over the surface of an ice crystal. Similar to air, water molecules migrate from an area of high pressure to an area where pres-

Precipitation

sure is lower. Thus, when a water droplet comes in contact with an ice crystal, water molecules leave the water droplet and freeze onto the ice crystal. The greater the difference in vapor pressure between water and ice, the more rapidly ice crystal growth occurs.

The size of a snowflake depends upon the temperature of the air as the snowflake descends. When a snowflake falls through air in which the temperature is above freezing, it melts around the edges. A film of water forms that acts like glue. It causes snowflakes that come in contact with one another to stick together, producing large, soggy snowflakes, 2 to 4 inches (5 to 10 centimeters) or larger in diameter. These snowflakes stick to surfaces and are heavy to shovel.

In contrast, snowflakes that fall through very cold, dry air, do not readily stick together. They are small and powdery when they hit the ground. This snow makes for ideal skiing conditions.

SNOW GRAINS AND SNOW PELLETS

Snow grains are the frozen equivalent of **drizzle.** They are small, soft, white grains of ice that form within **stratus** clouds and fall to the ground in only small amounts. Snow grains are elongated and generally have diameters less than .04 inches (.1 centimeters). Because they are so light, they fall very slowly and land gently, without bouncing or shattering.

In contrast to snow grains, **snow pellets** do fall rapidly and bounce high off the ground. These white pieces of icy matter, also called **graupel** or **soft hail,** measure between .08 and .2 inches (.2 and .5 centimeters) in diameter. Snow pellets fall in showers and feel brittle and crunchy underfoot.

Snow pellets form within towering **cumuliform** clouds, where the atmosphere is very unstable. This instability occurs when air temperature drops rapidly with height. The top of the cloud, where temperatures are lowest, is inhabited mostly by ice crystals. The ice crystals grow by the deposition of water vapor onto them, and take on the shape of snowflakes.

As a snowflake travels downward into the warmer, middle region of the cloud, it encounters supercooled water droplets. In what is known as **riming,** the droplets freeze to the snowflake, trapping numerous air pockets in the process. If riming occurs to a great enough extent, the snowflake will be transformed into a lumpy, white pellet of snow called graupel. It is in this form that precipitation reaches the surface.

INTENSITY OF SNOWFALL

The lightest form of snowfall is flurries. Flurries are brief and intermittent and originate in cumuliform clouds. While they produce very little accumulation, flurries may interfere with visibility.

A heavier and more persistent snowfall, which most people think of as "snow," comes from **nimbostratus** and **altostratus** clouds. This snowfall may continue steadily for several hours.

A brief but heavy snow shower is called a **snow squall.** Snow squalls, like flurries usually originate in cumuliform clouds. Snow squalls can be compared in intensity to summer rain showers and are accompanied by strong surface winds.

Heavy snow is defined as that which reduces visibility to .3 miles (.5 kilometers). Heavy snow, on average, yields 4 inches (10 centimeters) or more in a twelve-hour period or 6 inches (15 centimeters) or more in a twenty-four hour period. However, the amount of snow accumulation deemed "heavy" varies from one geographic area to another. For instance, in places where accumulations of 4 inches during a twelve-hour period are common, snow may not be considered "heavy" until over 6 inches have accumulated during that period. On the other hand, where any accumulation of snowfall is rare, an accumulation of 2 to 3 inches (5 to 8 centimeters) in a twelve-hour period may be considered "heavy."

A heavy snowfall.

Precipitation

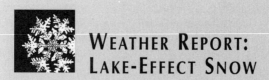

WEATHER REPORT: LAKE-EFFECT SNOW

Lake-effect snow is the name given to the heavy snowfalls that occur along the shorelines of the Great Lakes. The process that gives rise to lake-effect snow begins when dry, polar air masses, informally called **Alberta Clippers,** sweep down from Canada. As this air travels across the Great Lakes, evaporation in the form of **steam fog** raises the humidity of the air considerably. Clouds form and deposit heavy snowfall on the land downwind of the Great Lakes.

In January 1959, 51 inches (130 centimeters) of lake-effect snow fell during a sixteen-hour period on Bennets Bridge, New York, on the southern shore of Lake Ontario. And lake-effect snow fell on Buffalo, New York, for forty straight days during the winter of 1976–77.

Snow can also be classified by how it behaves on the surface. For instance, **drifting snow** is loose snow that has been swept into large piles, or "drifts," by a strong winds. **Blowing snow** is snow that has been lifted off the surface by the wind and blown about in the air. Blowing snow may reduce visibility in a manner similar to that which occurs in a heavy snowfall. A **ground blizzard** is the condition that results when snow continues to drift and blow after a snowfall has ended.

Fall streaks are snow that never reaches the ground. Similar to **virga,** they are not considered true precipitation. While virga starts out as rain that evaporates into dry air as it falls, fall streaks start out as ice crystals that **sublimate** into dry air. Fall streaks typically come from **cirrus** clouds. They hang beneath the cloud like white streamers, blown horizontally by the wind.

BLIZZARDS

A blizzard is the most severe type of winter storm. It is characterized by strong winds, large quantities of snow, and low temperatures. The National Weather Service defines a blizzard as a snowstorm with winds of 35 mph (56 kph) or greater. The temperature is generally 20°F (-6°C) or lower. The falling and blowing of fine, powdery snow greatly reduces visibility, often to less than a quarter of a mile and sometimes to just a few yards.

When a blizzard strikes, it can bring traffic to a standstill, strand motorists, and shut down entire cities. Prolonged exposure to a blizzard can cause **frostbite, hypothermia,** and even death (see "Temperature Extremes, Floods, and Droughts," page 303). The cause of some deaths in blizzards has been suffocation; people have actually choked to death on fine, powdery snow.

A **severe blizzard** is a blizzard in which wind speeds exceed 45 mph (72 kph), the snowfall is heavy, and the temperature is no higher than 10°F (-12°C). When falling, drifting, and blowing snow reduce visibility to almost zero, the condition is called a **whiteout.** Everything appears white, making the ground and sky indistinguishable. It's easy for someone who is stranded in such conditions to become disoriented and lose their way.

AVALANCHES

An **avalanche** is the cascading of at least 100,000 tons of snow down a steep slope. It occurs when stress is placed on a weak layer of snow. For an avalanche to occur, snow on the ground must be layered in such a way that it is structurally unstable. For instance, a loose layer of snow may be sandwiched between two more compact layers.

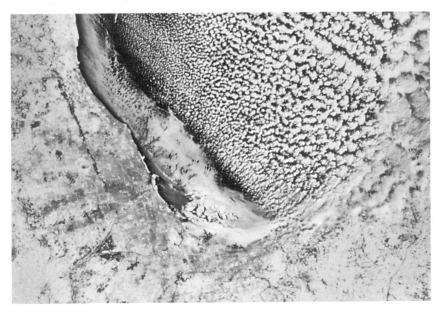

A Landsat image shows lines of intense lake-effect snow squalls heading for northern Indiana and lower Michigan during a blustery outbreak of frigid Arctic air moving across Lake Michigan.

Precipitation

A Key Reference To: Winter Storm Alerts and Safety Procedures

The National Weather Service issues winter storm alerts whenever snowfall is anticipated to be heavy enough to create dangerous travel conditions. There are four different types of these alerts based on the seriousness of the storm, defined as follows:

- A winter weather advisory states that snow, sleet, **freezing rain,** or high winds may be on the way. It advises people to exercise caution when traveling.
- A winter storm **watch** states that at least 6 inches of snow and an **ice storm** may be on the way. It advises people to limit their travels and to exercise great caution if they must venture onto the roads.
- A winter storm **warning** states that a storm, including **heavy snow** and possibly ice, has already begun or will soon begin. It advises people not to travel, except in an emergency.
- A **blizzard** warning states that a blizzard—including heavy snow, low temperatures, and winds of at least 35 mph (56 kph)—is on the way. The combination of heavy snowfall and low clouds make it appear that the ground and sky are continuous white sheet (called a **whiteout**), making travel nearly impossible. It advises people to remain indoors.

Safety procedures:

If you live in an area affected by winter storms, it is wise to take the following precautions at the start of the season:

- Store extra blankets and warm clothing and boots at home for every member of the family.
- Put together a supply kit for your home containing first aid materials, a battery-powered flashlight, a battery-powered radio, extra batteries, non-perishable food, a non-electric can opener, and bottled water.
- Store a similar supply kit, plus the following equipment, in the trunk of your car: a shovel, a bag of sand, tire chains, jumper cables, and a piece of brightly colored cloth to tie to your antenna.

- Keep your car's gas tank full to prevent the fuel line from freezing.

If you must go outside during a winter storm, follow these rules:
- Wear several layers of lightweight clothing, gloves, a hat, and a scarf covering your mouth.
- Walk carefully over icy ground.
- When shoveling, take frequent breaks to avoid overexertion.
- If you must drive, inform someone of your route, destination, and expected time of arrival.

If you get stranded in your car during a winter storm:
- Stay with your car. Tie the brightly colored cloth to your antenna so rescuers can spot you. Don't attempt to walk away! It's easy to become disoriented and lose your way in a snowstorm.
- Only start the car and turn on the heater for ten minutes out of every hour. When the car is running, leave on the inside light so you can be spotted. When the car is not running, periodically check the tailpipe and clear it of snow, if necessary. If your tailpipe is blocked, dangerous fumes can back up into the car.
- Move your arms and legs continuously to stay warm and maintain your blood circulation.
- Let in fresh air by slightly opening the window that's opposite the direction of the blowing wind.

Every time new snow falls, it places additional weight and pressure on the existing snow. At some point, this pressure may give rise to a fracture (break) across the blanket of snow, down to the weakest layer. As soon as any additional stress is added, a slab of snow breaks off and goes hurtling down the slope, The additional stress may take the form of more snow, a strong gust of wind, the weight of a skier, or even a loud noise.

Avalanches are extremely destructive. They bury everything in their path, even cities, in a matter of seconds. The largest avalanches occur in the Andes, the Himalayas, and the mountains of Alaska. However it is in the Alps, where valley regions are heavily populated, that avalanches pose the greatest danger to humans.

Precipitation

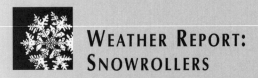

WEATHER REPORT: SNOWROLLERS

On rare occasions, Mother Nature gives us a hand in building a snowman by creating **snowroller**s. A snowroller is a lumpy, spherical or cylindrical mass of snow, generally less than 1 foot in diameter. It is created only under a very particular set of conditions. First, there must be a layer of smooth, hard, crusty old snow on the ground. Then, a light layer of new snow falls on top of the old snow. Finally, a strong, warm wind blows in, rapidly raising the temperature.

This wind literally lifts up a patch of snow and rolls it along the surface snow, which is warm and sticky. This process continues until the accumulated snow becomes too heavy to be rolled any further.

One place where snowrollers have been witnessed is in Boulder, Colorado. They are created by the strong, warm **chinook** winds (see "Local Winds," page 128) that blow down the slopes of the Rocky Mountains.

For more information on snowrollers, see: Schlatter, Thomas. "Weather Queries: Snowrollers." *Weatherwise.* Dec. 1996/Jan. 1997: 42.

In the United States, between twelve hundred and eighteen hundred avalanches are reported each year. Most of these occur in the western states. When you take into account avalanches that go unnoticed or unreported, the actual number is much higher.

Colorado has the highest death rate caused by avalanches, with six to eight fatalities a year. Most of the people killed are skiers or snowmobilers. The best way to stay out of an avalanche's path is to avoid snow-covered slopes at angles steeper than 30 degrees.

ICE

This section covers **ice pellets** and **hailstones,** the two forms of precipitation that fall to the ground as hard, mostly transparent pieces of

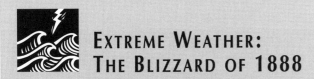

Precipitation

EXTREME WEATHER: THE BLIZZARD OF 1888

In March 1888, a blizzard hit the East Coast and set snowfall records from Virginia to Maine, many of which remain to this day. New York City received 21 inches (53 centimeters) of snow in about twelve hours. Power lines were brought to the ground, halting the city's new telephone system, streetcars, and elevated trains. It was during that blizzard that the New York Stock Exchange closed for the first time since the Stock Exchange's opening in 1790.

All throughout southern New England and southeastern New York, snowfalls averaged 40 inches (100 centimeters) or more and winds were recorded at 50 to 70 mph (80 to 112 kph). The largest snowfall was recorded at Saratoga Springs, New York, with 58 inches (147 centimeters). Gravesend, New York, had the highest reported snow drift: 52 feet (16 meters)!

Four hundred people perished in the blizzard, half of them in New York City.

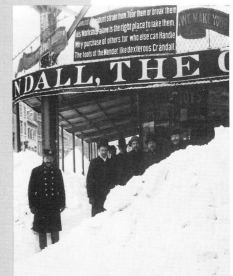

ice. As you will see, ice pellets and hailstones have little else in common besides their composition.

ICE PELLETS

Ice pellets are frozen raindrops. They are formed by precipitation that passes first through a warm layer of air and melts, then re-enters a layer of freezing air and re-freezes. The precipitation reaches the ground as tiny pellets of ice.

Ice pellets differ from **freezing rain** in the depth of the layer of freezing air through which they pass. Ice pellets encounter the freezing air at a higher elevation than does freezing rain. Thus, ice pellets have

Precipitation

time to freeze *before* they hit the ground, while freezing rain freezes only *on contact* with the cold ground.

In the United States, ice pellets are also referred to as "sleet." Sleet is a confusing term, however, since this word is used in Australia and Great Britain to refer to a mixture of rain and wet snow. Even in the United States, sleet is often used by the news media to describe slushy precipitation. For this reason, we will use the term "ice pellets" instead of "sleet" in this text.

Ice pellets measure only .2 inches (.5 centimeters) in diameter, are irregular in shape, and bounce when they hit a surface. Ice pellets can also be identified by the "ping-ping" sound they make when they strike a glass or metal surface. An accumulation of ice pellets can create hazardous driving and walking conditions.

HAILSTONES

Hailstones are a larger and potentially much more destructive form of frozen precipitation than are ice pellets. They have either a smooth or jagged surface and are either totally or partially transparent. While most hailstones are pea-sized, they may reach the size of softballs.

Large hailstones have been responsible for destroying crops, breaking windows, and denting cars. They have also caused numerous human and animal deaths. The largest single hailstone on record was about the

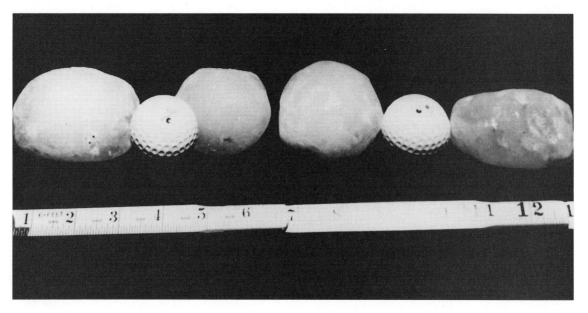

Hailstones that fell to the ground in Texas are compared to golf balls.

WEATHER REPORT: HAIL ALLEY

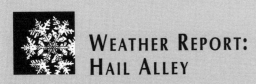

"Hail alley" is the region of North America where hail frequently damages crops. It covers a north-south belt from Alberta, Canada, to Texas. It extends westward to the Rockies and eastward to the Mississippi River. Hailstorms occur with the greatest frequency in the Great Plains states. There, hail falls accompany 10 percent of all thunderstorms.

A hailstorm can flatten an entire field in minutes. Hail damage to crops in the United States alone tops $700 million per year. To guard against the chance that a single storm can wipe out an entire year's earnings, farmers in "hail alley" spend large sums on hail insurance. Illinois farmers top the list, purchasing over $600 million worth of liability coverage annually.

size of a softball. It measured 5.5 inches (14 centimeters) in diameter and 17 inches (44 centimeters) in circumference, with a weight of 1.7 pounds (3.7 kilograms). It fell on Coffeyville, Kansas, on September 3, 1970.

Hailstones are formed within **cumulonimbus** clouds during intense **thunderstorm**s. A hailstone forms around a small particle, called an "embryo." Objects that can act as embryos include ice crystals, frozen raindrops, **graupel,** dirt, or even insects. There have also been reports of larger organisms, such as frogs, being swept up in a **tornado** and returning to Earth with a hailstone formed around them.

As an embryo travels through a cloud, it is coated by cloud droplets and grows larger by **accretion** (gradual accumulation). On reaching the bottom of the cloud, the developing hailstone is blown back up to the top by powerful **updraft**s. It repeats its journey down and up through the cloud many times. In the case of very strong updrafts, this process may last several minutes. When the hailstone becomes heavy enough to overcome the force of the updraft, it falls to the ground.

Thunderstorms, and hence hailstones, are warm-weather phenomena. As a hailstone descends toward the surface and encounters warm air,

Precipitation

WEATHER REPORT: HAIL SUPPRESSION

Because of the extensive damage caused by large **hailstones,** people from many countries have attempted, historically and to this day, to prevent the large balls of ice from falling. In ancient times, people shot arrows into the clouds or rang church bells to frighten away evil "hail" spirits. European farmers in the sixteenth through nineteenth centuries fired rockets and cannons toward the clouds. Today's method of hail suppression involves seeding clouds with silver iodide or lead iodide, similar to the method used to increase **precipitation** (see Rainmaking box, pages 190–91). It's impossible to say with certainty whether today's high-tech methods are achieving a success rate any higher than that of primitive methods.

The goal of modern hail suppression practices is to keep hailstones from reaching destructive sizes. Like fog dispersal, hail suppression operates on the principle that the greater the number of particles, or **freezing nuclei,** there are in a cloud, the smaller each nucleus will remain. This is because there is a finite number of **supercooled water** droplets within a cloud and the freezing nuclei compete for those droplets. Thus, the growth of each nucleus is limited.

The results of hail suppression efforts, at present, are inconclusive. Russian scientists have claimed great success, but it has not been possible to verify their results. Scientists in the United States, France, Switzerland, Italy, and Canada may have also achieved some degree of success. Experimenters in the United States believe that it's theoretically possible for hail suppression techniques to bring about a 20 to 40 percent reduction in damaging hail. Hail suppression currently faces opposition from those who believe that the practice may also reduce rainfall.

it begins to melt. Small hailstones may melt completely in the air. Hailstones that are large enough, however, melt only partially and reach the ground in the frozen state. In the tropics, where the air is very warm, hail always melts before reaching the ground, turning to rain.

If you slice a hailstone in half, you will notice that it resembles an onion, with a pattern of concentric rings. The number of rings is equal to the number of trips the hailstone made through the cloud. Up to twenty-five rings have been counted in large hailstones.

You will also notice that the layers of a hailstone alternate between clear ice and white ice, called **rime.** The clear layers are formed when the hailstone is in warmer air, in the lower portion of the cloud. There, **supercooled water** droplets are plentiful. They form a layer of water around the hailstone, which slowly freezes when the hailstone returns to the colder, upper portion of the cloud.

The milky white layers are formed in the upper, freezing portion of the cloud, where supercooled droplets are scarcer and freeze directly onto the hailstone by the process of **riming.** Similar to the process by which a graupel is formed, the droplets trap air bubbles when they freeze to the hailstone.

The accumulation of hailstones during a thunderstorm can be considerable. One of the largest **hail** falls on record reached a depth of 18 inches (46 centimeters) and occurred in Selden, Kansas, in June 1959. Occasionally, snowplows must be taken out of summer storage to clear roads of hailstones. This was the case in September 1988, in Milwaukee, Wisconsin, when hailstone drifts reached 18 inches. And in August 1980, snowplows were required to removed hailstone drifts 6 feet (2 meters) deep in Orient, Iowa.

The hailstorms that have caused the greatest toll in terms of human life have occurred in Asia. In 1888, in northern India, hail killed two hundred forty-six people. In 1932, in southeast China, a hailstorm claimed about two hundred lives and injured thousands. And in 1986, in Bangladesh, a storm that produced some very unusual hailstones weighing more than 2 pounds (1 kilogram) each, killed ninety-two people. In the United States, only two deaths have been attributed to hail in the last century.

If you are caught in a thunderstorm, watch for these warning signs of hail: a green tinge develops at the base of the cloud or the rain begins

Precipitation

to take on a whitish color. If you witness either of these signs, it is time to collect your family and pets and move indoors.

6

THUNDERSTORMS

Thunderstorms are relatively small, but intense, storm systems. An average thunderstorm is only 15 miles (24 kilometers) in diameter and lasts about thirty minutes. During that time, it produces strong winds, heavy rain, and **lightning.** Only 10 percent of thunderstorms are considered "severe," meaning they produce some combination of high winds, **hail, flash flood**s, and **tornado**es. Tornadoes are spawned by only about 1 percent of all thunderstorms.

Severe thunderstorms and their related phenomena produce significant human injuries and fatalities, as well as property damage, each year. Hailstorms, for instance, are responsible for nearly $1 billion a year in crop damage. And tornadoes cause about eighty deaths and fifteen hundred injuries annually. While straight-line winds called **derecho**s occur less frequently and result in fewer fatalities, a single derecho can cause millions of dollars in damage.

Lightning, which occurs with all thunderstorms, sets off about ten thousand forest fires each year in the United States alone, causing several hundred million dollars in property damage. Furthermore, lightning causes between 75 and 100 deaths and about 550 injuries annually in the United States. It is the second biggest weather killer in the country, topped only by flash floods. Flash floods, which are sudden, intense, localized floodings caused by heavy rainfall, kill an average of 140 people every year.

The **cumulonimbus** clouds that produce thunderstorms are giant storehouses of energy. A typical thunderstorm unleashes 125 million gallons of water and enough electricity to provide power to the entire United States for twenty minutes.

Thunderstorms

At any given time, there are about two thousand thunderstorms underway around the world. About forty thousand thunderstorms occur each day and fourteen million thunderstorms each year, worldwide. Earth is struck by lightning from these storms 100 times every second.

In this chapter we will examine thunderstorms: their development and behavior; the various types; the secondary weather effects they produce; and the safety procedures to be undertaken during thunderstorms and flash floods.

Watching a thunderstorm approach can be quite an exhilarating experience. Imagine standing on your front porch on a hot, humid afternoon. The morning's **haze** has given way to a line of tall **cumulus** clouds. On the horizon there are enormous **thunderstorm clouds,** with their whitish tops and dark undersides.

As these clouds approach, the sky darkens, almost blocking out the sunlight. Then comes a calm period in which the air feels still, hot and very muggy. Next the wind picks up and the rain begins to fall in large drops. Soon the rain intensifies to a downpour and the wind turns cold and blows wildly. Lightning brightens the sky here and there. Then it strikes nearby and is followed by a thunderclap so loud it jolts you to your feet.

Half an hour later the storm is over. The storm clouds move away, the sun comes out, and the air is cooler and less humid. Before long, however, the heat returns.

EVOLUTION OF THUNDERSTORMS

Two atmospheric conditions are required for the development of a thunderstorm. The first is that the surface air must be warm and humid. The other is that the atmosphere must be **unstable.** As we learned in the chapter "What Is Weather?" "unstable" means that the surrounding air is colder than a rising **air parcel.** As long as the atmosphere remains unstable, an air parcel will continue to rise. When the air parcel reaches a height at which the atmosphere is **stable,** meaning that the surrounding air is warmer than the air parcel, it will rise no further.

Thunderstorms occur when warm, moist air rises quickly through an unstable atmosphere. On reaching the **dew point,** moisture within the air condenses, forming a cloud. If the atmospheric instability is great enough, this air will rise to great heights, and the cloud will develop ver-

*Opposite page:
Figure 20: The life cycle
of a thunderstorm.*

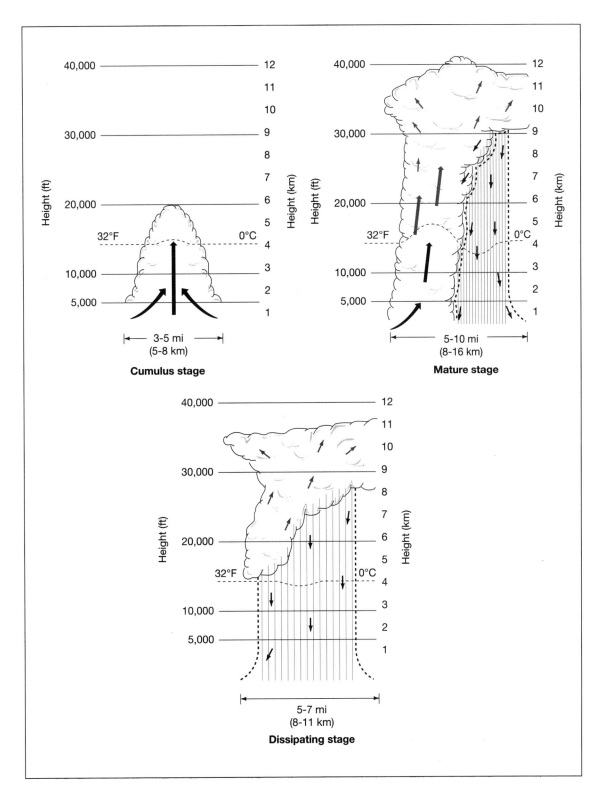

Thunderstorms

tically into a towering cumulus cloud. In conditions of great instability, the cloud will develop into a full-fledged cumulonimbus, or thunderstorm, cloud.

A number of factors may trigger the rising of warm air, for example: an **air mass** rides up, along a mountainside; air is forced upward by an advancing **cold front**; and/or **thermal**s, pockets of rising warm air, are produced by uneven heating of the surface. At the same time, there must be a **divergence** of **winds aloft.** This divergence causes surface winds to **converge** beneath, and rise to, the point of divergence. (For more information on convergence and divergence, see "What Is Weather?" on page 32.)

The life cycle of a thunderstorm can be broken down into three stages which trace its development from inception to dispersion. These stages, depicted in Figure 20, are called: the **cumulus stage,** the **mature stage,** and the **dissipating stage.**

Cumulus Stage

A thunderstorm begins its development in the cumulus stage, also known as the **developing stage.** A thunderstorm usually begins forming late in the afternoon or early in the evening. It follows a period in which cumulus clouds have been forming, then evaporating into the dry air, only to form again at higher altitudes. Each time the clouds evaporate, they raise the humidity of the air. This fact is important because as long as the air is dry, any moisture that condenses within rising warm air will quickly evaporate. Only when the air is humid will moisture condense into a cloud that remains in the air. As the air is humidified at progressively higher levels, the stage is set for the development of towering cumulus clouds.

In the cumulus stage, which takes only fifteen minutes or so, cumulus clouds undergo dramatic vertical growth. The cloud tops rise to a height of about 30,000 feet (9,100 meters). At the same time, the clouds spread horizontally and merge into a line up to about 9 miles (15 kilometers) across.

As described in the chapter "What Is Weather?" when air rises it cools by the **dry adiabatic lapse rate** (5.5°F per 1,000 feet). Once the air has cooled to the dew point, it becomes **saturated** and the moisture within it condenses. **Condensation** releases **latent heat** into the cloud. This heat increases the temperature contrast between the cloud and the surrounding air, fueling the upward growth of the cloud, as long as warm air continues rising into it.

Furthermore, once air enters a cloud and becomes saturated, it cools by the **moist adiabatic lapse rate** (3.3°F per 1,000 feet). Thus, it cools more slowly than it did at the dry adiabatic lapse rate, when it was **unsaturated.** This change enables the air to rise to even greater heights before reaching a stable layer of atmosphere.

Air will continue rising as long as it is warmer, and less dense, than the surrounding air. The greater the instability of the air (the more rapidly the air cools with height), the higher the air parcel will ascend.

As air rises beyond the top of the cloud and into the dry air, the cycle is repeated. The moisture evaporates and increases the humidity of the dry air, enabling condensation to take place at ever-greater altitudes. In this way, a **cumulus** cloud develops upward, into a **cumulonimbus** cloud.

Precipitation rarely occurs during the cumulus stage, because water droplets or ice crystals are blown upward by rising air, into the tops of the clouds. Within the cloud, the **updraft**s, columns of air blowing upward, may exceed 30 feet per second (10 meters per second). Lightning and **thunder** are produced only occasionally during the cumulus stage.

Mature Stage

The **mature stage** of a thunderstorm begins when the first drops of rain reach the ground. It is during the mature stage that one sees **heavy rain,** strong winds, lightning, and sometimes hail and tornadoes. If the thunderstorm is severe, the sky may appear black or dark green. A thunderstorm generally remains in the mature stage for ten to thirty minutes, occasionally longer.

Throughout the mature stage, the cumulonimbus cloud continues building. Eventually it builds to the **tropopause,** which is between 5 and 7 miles (8 to 11 kilometers) above Earth's surface, and may even overshoot the tropopause. At that point, the rising air encounters a stable atmosphere. It ceases to rise and spreads out laterally in the **anvil** shape that characterizes the top of mature thunderstorm clouds. The base of cloud, meanwhile, grows to several miles across.

Precipitation begins to fall from the thunderstorm cloud when ice crystals or water drops within the cloud reach a critical mass. That is, they become large enough to overcome the updrafts that have previously confined them to the tops of the clouds. As the precipitation falls, it pulls air with it. These downward blasts of air, felt at the surface as cool gusts, are called **downdraft**s.

Thunderstorms

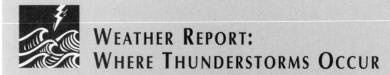

WEATHER REPORT: WHERE THUNDERSTORMS OCCUR

Over fourteen million thunderstorms take place throughout the world, annually. For the most part, they occur in warm, humid areas. The world's greatest concentrations of thunderstorms are in Brazil's Amazon Basin, the Congo Basin of equatorial Africa, and in the islands of Indonesia. In each of these areas, thunderstorms occur on more than 100 days each year.

Thunderstorms occur, with varying frequency, all throughout the United States. Florida's Gulf Coast has thunderstorms more often than any other U.S. location. Thunderstorms occur there on 130 days per year, on average. On the other extreme is the Pacific Coast, which sees thunderstorms on only 5 to 10 days per year, and Alaska, which only has one thunderstorm every three to five years.

In between those two extremes are the following annual averages: 1) the rest of Florida's Gulf Coast, plus the Gulf Coasts of Alabama and Mississippi, have thunderstorms on 80–100 days; 2) the rest of the southeastern United States has thunderstorms on 60–80 days; 3) the central portion of the Rockies has thunderstorms on 50–70 days; 4) the Corn Belt (Iowa, Indiana, and Illinois) and Great Plains states (states just east of the Rockies) have thunderstorms on about 50 days; 5) the portion of the Midwest that lies east of Iowa, as well as the mid-Atlantic states and New England, has thunderstorms on 20–40 days.

The updrafts of air continue to bring warm, humid air into the thunderstorm throughout the mature stage. These updrafts create a situation in which there are columns of rising air adjacent to columns of descending air. The rising air builds up the storm cloud while the descending air returns the cloud's moisture to Earth.

As the storm progresses, the updrafts weaken and downdrafts strengthen by **entrainment.** Entrainment is the process by which cool, unsaturated air next to a cloud gets pulled into the cloud. As this dry air mixes with air in the cloud, the **relative humidity** in the cloud is lowered. Some of the water droplets evaporate, a process that absorbs latent heat from the cloud. This has the opposite effect that condensation had in the cumulus stage. Namely, **evaporation** cools the rising air and slows its ascent. It also cools the downdrafts. Since cold air is denser and heavier than warm air, this cooling causes the downdrafts to fall faster.

The thunderstorm peaks in intensity at the end of the mature stage. As downdrafts begin to dominate updrafts, the thunderstorm yields the heaviest rain, the most frequent lightning, and the strongest winds. If the thunderstorm produces hail or tornadoes, they will occur at the end of the mature stage.

Dissipating Stage

In the **dissipating stage** of a thunderstorm, precipitation falls from the entire cloud base. Downdrafts overtake updrafts, preventing warm, moist air from rising up into the cloud. In effect, the downdrafts cut off

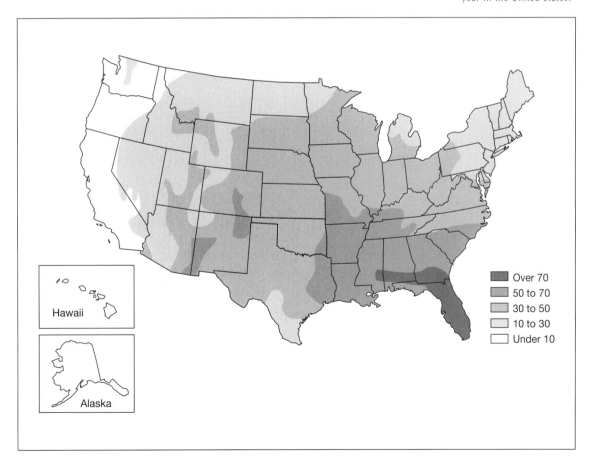

Figure 21: Average number of days on which thunderstorms occur each year in the United States.

Thunderstorms

the thunderstorm's fuel supply. Without a constant influx of moisture from below, the cloud begins to evaporate.

During the dissipating stage, rain becomes light and winds become weak. An hour or so after the cumulus stage began, the storm cloud dissipates, leaving only wispy traces of **cirrus** high in the sky. The cool, refreshing air brought by downdrafts during the mature stage subsides. The rain evaporates, which increases the humidity of the air. The heat and humidity following the thunderstorm may create even more oppressive conditions than what existed before the thunderstorm.

The preceding explanation of a thunderstorm's life cycle applies to a single **convective cell** of a thunderstorm. A convective cell is a unit within a thunderstorm cloud that contains updrafts and downdrafts. While some thunderstorms are of the single-cell variety, most thunderstorms contain several convective cells and are called **multicell thunderstorms.**

In the case of a multicell thunderstorm, convective cells are simultaneously in various stages of development. Old cells die and new cells form as the storm moves over the ground and encounters fresh sources of warm, moist air. The life cycle of each convective cell lasts thirty minutes to sixty minutes. A multicell thunderstorm may last several hours.

TYPES OF THUNDERSTORMS

Thunderstorms are classified by a number of criteria. The first criterion is the mechanism that triggers its formation. Another criterion is whether the thunderstorm is isolated or part of a cluster of thunderstorms. Finally, thunderstorms are classified on the basis of their severity. These groupings often overlap. For instance, a thunderstorm that forms along a **cold front** may be weak or severe, and may occur singly or in a line of thunderstorms. In this section, we will examine the origin of thunderstorms and the ways in which specific types of thunderstorms behave.

AIR MASS THUNDERSTORMS

An **air mass thunderstorm,** the most common type of thunderstorm, is one that forms within a single mass of warm, humid air. **Air mass** thunderstorms are relatively weak, meaning they don't produce **hail** or strong winds, and die out quickly. They do, however, produce **lightning** and sometimes destructive, downward gusts of wind called **downburst**s. Air mass thunderstorms most often form in the late afternoon, at the warmest time of day. In regions outside of the tropics, they form mostly during spring and summer months.

For any thunderstorm to develop, air must be lifted to the level at which it is **saturated**, meaning the moisture within it condenses and forms a cloud. The air does not rise to its condensation level automatically. Rather, it requires a lifting mechanism. In air mass thunderstorms, that lifting mechanism is the intense heating of small areas on the surface, which produces rising pockets of air called **thermal**s. This heating is most often accompanied by a **convergence** of surface winds and a resultant uplift of air. The air ascends to a point where there is a **divergence** of **winds aloft.**

An air mass thunderstorm occurs in isolation from other thunderstorms. It may be composed of a single **convective cell.** More often, however, it contains multiple cells.

Air mass thunderstorms are common in the central and eastern United States in spring and summer. They are initiated by the northward flow of humid, tropical air masses from the Gulf of Mexico, the Caribbean, and the Atlantic Ocean near Bermuda. On an average, these thunderstorms occur one afternoon out of every three on the Gulf Coast of Florida.

Orographic Thunderstorms

An **orographic thunderstorm,** also called a mountain thunderstorm, is a type of air-mass thunderstorm that is initiated by the flow of warm air up, along a mountainside. Such storms occur most commonly on slopes with greatest exposure to the sun.

As a slope is heated, the air adjacent to it is also heated. That warm air rises, cools and the moisture within it condenses to form **cumulus** clouds. Given a constant influx of warm, moist air and an unstable atmosphere, these clouds will develop into **cumulonimbus** clouds.

Orographic thunderstorms are relatively weak. They are common in the Rocky Mountains, as afternoon breezes lift warm air up the mountainsides. Because of this lifting mechanism, mountainous regions in the United States are hit with more thunderstorms than any region outside southeastern states.

Frontal Thunderstorms

Frontal thunderstorms form along the edge of a **front.** They occur most often when a cold front is displacing a maritime tropical (warm, moist) air mass which has remained stationary for several days. As explained in the chapter "What Is Weather?" (see page 39) an advanc-

Thunderstorms

ing cold front wedges underneath an existing warm air mass, thrusting the warm air upward.

Less frequently, a frontal thunderstorm is initiated by an advancing **warm front.** Such storms occur only if advancing warm air, which glides up and over the residing cold air mass, is entering a particularly **unstable air layer.**

Frontal thunderstorms may occur any time of day or night and at any time of year, except for very cold winter days. They are most likely to form in warm weather, when **convection** is enhanced by heating of the ground. Frontal thunderstorms that occur in the winter may yield snow and are generally far weaker than their summertime counterparts.

Though they often have stronger wind and heavier rain than air mass thunderstorms, frontal thunderstorms are relatively weak systems. Under certain conditions, thunderstorms produced along a front are severe. This is the case when a series of thunderstorms, called a **squall line,** arises in a band running parallel to the front. We will return to squall lines in our discussion of "severe thunderstorms."

Frontal thunderstorms are most common in the Great Plains states and the Midwest, where cold fronts from Canada overtake warm, moist air from the Gulf of Mexico. More than half the world's **tornado**es occur in this region.

MESOSCALE CONVECTIVE COMPLEXES

A **mesoscale convective complex** (MCC) is a group of thunderstorms that forms a nearly circular pattern over an area that is about one thousand times the size of an individual thunderstorm. An MCC may cover an area greater than 50,000 square miles (130,000 square kilometers), or the size of a small state.

The combined effect of the individual thunderstorms of the MCC is to produce an airflow that favors the formation of new thunderstorms.

Thunderstorms continually form and dissipate within an MCC. The overall pattern persists for up to twenty-four hours and moves very slowly, usually less than 20 mph (30 kph). MCCs yield significant amounts of rainfall. In the Great Plains states and the Midwest, MCCs produce around 80 percent of the rainfall during the growing season. Around half of all MCCs are severe, spawning any combination of tornadoes, **flash flood**s, hailstorms, and high winds.

MCCs most often arise in warm weather and at night. More than fifty MCCs per year form over the central and eastern United States.

SEVERE THUNDERSTORMS

The National Weather Service defines a thunderstorm as "severe" if it has one or more of the following elements: wind gusts of at least 58 mph (93 kph); **hailstone**s at least 3/4 inch (2 centimeters) in diameter; or tornadoes or **funnel cloud**s. **Severe thunderstorm**s may also be accompanied by flash floods.

Severe thunderstorms are formed in the same way as more moderate thunderstorms: by the rising of moist air into an unstable atmosphere. A strong cold front is most often the force that provides the vigorous uplift of warm air required to produce a severe thunderstorm. At the same time, the moist surface air is pulled upward when a **divergence,** or flow away from a central point, occurs in the **winds aloft.** This divergence triggers the convergence of surface winds beneath that point. The surface winds then rise to the area of divergence above.

One condition that gives rise to some of the largest and most severe thunderstorms is that an **inversion** is present for much of the day. An inversion is the increase of air temperature with height, through some portion of the atmosphere. The presence of a warm air layer aloft acts as a lid that prevents warm, humid surface air from rising. In other words, an inversion produces an absolutely **stable** atmosphere. As a result, only shallow cumulus clouds can form.

Sometimes on a summer day during which an inversion has occurred, surface air will become heated to the point at which it is warmer than the warm air aloft. Pockets of warm air will then burst through the warm layer, creating towering clouds with an explosive force. These clouds rapidly develop into severe thunderstorms.

An important factor in the formation of a severe thunderstorm is that **updraft**s are not weakened by falling **precipitation.** Such updrafts are produced in one of two ways. First, the updrafts are so strong that they keep all precipitation suspended in the cloud top for a long period, while the thunderstorm builds. Second, the updrafts are tilted so that precipitation falls alongside them, rather than into them. The updrafts become tilted by strong upper-level winds.

When the updrafts are tilted, precipitation falls into dry air alongside the updrafts, rather than directly into the updrafts.

Thunderstorms

Updrafts that are not weakened by falling precipitation are able to continue building the cloud top upward to greater and greater heights. Meanwhile, the precipitation falls into dry air adjacent to the updrafts and partially or completely evaporates. The dry air becomes cooler and denser and plunges downward.

The updrafts in a severe thunderstorm travel at speeds of 50 mph (80 kph) or greater. And they remain strong for far longer than they do in a weaker thunderstorm. Sometime the updrafts are so powerful that they rise above the **troposphere** and penetrate the **stratosphere.** This condition is called **overshooting.**

One effect of the strong updrafts in a severe thunderstorm is to keep hailstones suspended in the cloud for longer than usual. During that time the hailstones receive several coatings of ice and become quite large. When they become so heavy that they can not be supported by updrafts, the large hailstones either descend with a downdraft or are tossed through the side of the cloud by an updraft.

The **downdraft**s in a severe thunderstorm are also very strong. When strong, cool downdrafts reach the ground they further intensify the storm by displacing the warm, moist air and forcing it back up into the cloud. In this way, the storm is continually rejuvenated and can persist for several hours. Sometimes the warm, moist air that is forced upward has the effect of producing new thunderstorms. When strong downdrafts occur in the spreading **anvil** at the top the fully developed cloud, they may produce pouch-like **mammatus** projections on the anvil's underside.

The dividing line between cold downdrafts and warm air at the surface is called the **gust front.** Similar to a cold front, an advancing gust front is characterized by strong, shifting, and cold winds. The winds along a gust front can reach speeds of 55 mph (88 kph) or greater. In dry, dusty areas they carry debris along with them and create dust storms or sandstorms.

In some cases, a gust front can be clearly identified by the **roll cloud** that follows directly behind it. A roll cloud looks like a giant, elongated cylinder lying on its side that, as its name implies, is rolling forward. This cloud occupies a narrow vertical layer of air. The top of the cloud is prevented from developing upward by stable air at the base of the thunderstorm.

Another type of cloud associated with a gust front is a **shelf cloud.** A shelf cloud is fan-shaped with a flat base. It forms along the edge of the gust front as warm, humid air is thrust upward and encounters the stable air layer. In contrast to a roll cloud, which is a distinct formation, a shelf cloud is attached to the underside of the cumulonimbus cloud. Particularly violent winds blow on the surface beneath a shelf cloud.

SQUALL LINES. Most thunderstorms that are classified as "severe" exist in a band of thunderstorms called a **squall line.** A squall line may form either along a cold front or up to 200 miles (320 kilometers) in front of it. A squall line is particularly ominous in appearance. It looks like a churning, solid bank of fast-moving, low, dark clouds. A squall line may stretch for hundreds of miles. It moves along at speeds approaching 50 mph (80 kph).

As we described in the section on frontal thunderstorms, thunderstorms may form along a cold front when the cold front wedges beneath a warm, moist air mass. If the air mass being displaced is sufficiently moist, this upward thrust can cause vertical cloud development and thunderstorms.

When the squall line is ahead of the cold front, it is known as a **pre-frontal squall line.** Two processes may lead to the formation of a pre-frontal squall line. One process involves the lifting of warm, moist air by upper-level winds. It works like this: When upper level winds encounter a cold front, they flow over it. Once they have crossed the cold front, the upper-level winds then dip downward again. This sets in motion a wave-pattern of upper-level air flow. As the wave again flows upward, some 100 to 200 miles (160 to 320 kilometers) ahead of the cold front, it promotes the uplift of the warm, moist surface air.

A pre-frontal squall line may also form if the cold front is preceded by two air masses: a warm, dry air mass and a warm, moist air mass. In this case, thunderstorms don't form directly along the cold front, since the cold front is advancing on a dry air mass. However, the dry air mass is being pushed forward into the moist air mass, lifting the moist air upward. In that case, the squall line forms many miles ahead of the cold front.

SUPERCELL STORMS. **Supercell storm**s are the most destructive and long-lasting of all severe thunderstorms. They may continue for several hours and produce one strong tornado after another, as well as heavy rain and hail the size of golfballs. A supercell storm blazes a trail of destruction stretching 200 miles (320 kilometers) or more. For these reasons, the supercell has earned the title "The King of Thunderstorms."

A supercell storm arises from a single, powerful **convective cell.** It forms along cold front that is pushing its way through a mass of very

Thunderstorms

warm, humid air. A supercell storm may form in isolation or at the end of a squall line. Most supercell storms occur in spring and early summer, when temperature contrasts between warm and cold air masses is greatest.

The formation of a supercell thunderstorm requires a very specific arrangement of vertical air layers. In this explanation, we'll start with the surface layer of air and work upward. Note that some of the concepts involved in this explanation are explained in the chapter "What Is Weather?"

At the surface, the cold and warm air masses are rotating around a central area of low-pressure. At an altitude of about 5,000 feet (1,500 meters), above the surface low, there is a layer of warm, moist air blowing northward. Above that, at about 10,000 feet (3,000 meters) is a layer of cold, dry air, moving across from the southwest. This layer is called the **dry tongue.**

Located at the next highest layer, at about 18,000 feet (5,500 meters), are the **upper-air westerlies.** These winds aloft progress from west-to-east in a wave-like pattern of **ridges** and **troughs.** In this layer of air, the low-pressure center of a trough must be located just to the west of the warm surface air. Finally, at about 30,000 feet (9,000 meters) the **jet stream** produces an area of maximum speed where winds diverge (spread apart). At this level, surface air converges (comes together) and begins to rise.

Supercell thunderstorms on the prowl in Nebraska.

A supercell storm reaches immense proportions due to strong winds aloft. The motion of the upper-level winds tilts the storm so that updrafts remain free from falling precipitation. The precipitation falls downward into the dry air, creating downdrafts that force warm, moist air upward. The updrafts, in turn, continue adding fuel to the storm, causing the thunderstorm cloud to surge to tremendous heights.

Another necessary ingredient in the formation of a supercell is **wind shear.** Wind shear describes a condition in which vertical layers of wind increase in speed, and change direction, with height. When a layer of air is sandwiched between two other layers, each traveling at a different speed and direction, the sandwiched layer starts to roll (see Figure 23).

To illustrate how this occurs, take a chunk of clay and roll it between your palms. By moving your hands in opposite directions, you shape the clay into a long, snake-like column.

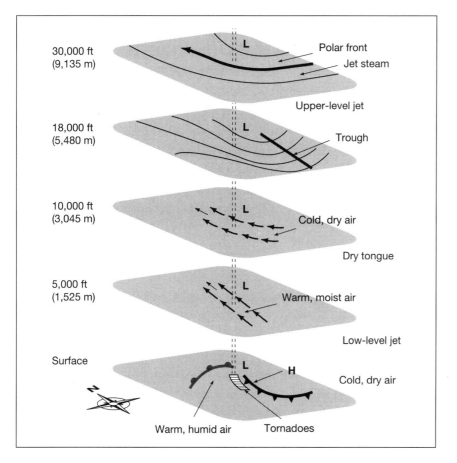

Figure 22: Arrangement of vertical air layers necessary for the formation of a supercell thunderstorm.

Thunderstorms

This demonstration shows what happens to a sandwiched air layer. It becomes a rotating column, like a rolling pin. That column is then turned upright, like a barber pole, by powerful updrafts. As updrafts continue to blow through the now vertical, spinning column, the updrafts themselves begin to rotate.

Wind shear creates a region of rotating updrafts within a supercell, called a **mesocyclone.** Mesocyclones are, on average, 10 miles (16 kilometers) in diameter although the diameters of the largest may reach 250 miles (400 kilometers). In addition to providing power to the thunderstorm, a mesocyclone is a necessary component in the formation of tornadoes. We will return to tornadoes in the next chapter.

THUNDERSTORM-ASSOCIATED PHENOMENA

As mentioned at the beginning of this chapter, a number of destructive and potentially deadly elements are associated with **thunderstorm**s. The most common of these are **lightning** and **thunder. Hail, flash flood**s, **tornado**es, and strong **downdraft**s of wind (including **macroburst**s, **microburst**s, and **derecho**s) are also well known to people who live in thunderstorm-prone areas. In this section we will discuss all of the aforementioned phenomena except hail, derechos, and tornadoes. To learn about hail and hailstorms, see "Precipitation," page 206. For an explanation of derechos, see "Local Winds," page 138.

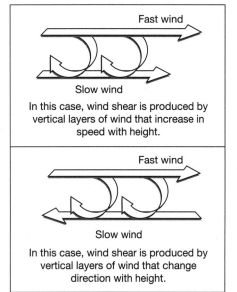

Figure 23: Wind shear.

LIGHTNING AND THUNDER

In order to qualify as a "thunderstorm," a rain shower or snow shower must be accompanied by lightning and thunder. Lightning is a short-lived, bright flash of light produced by a 100 million-volt electrical discharge. Lightning heats the air through which it travels to about 50,000°F (28,000°C). Compare this to the surface of the sun, which is about 11,000°F (6100°C)! Thunder is the sound wave that results when the intense heating causes the air to expand explosively.

At any given moment, approximately 100 lightning flashes are occurring worldwide. Lightning kills between 75 and 100 people in the United States each year and causes about 550 injuries. This is a greater number of deaths than those resulting from **hurricane**s or torna-

Thunderstorms

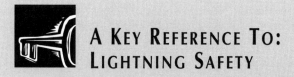

A Key Reference To: Lightning Safety

If you hear thunder, you are in the vicinity of lightning. Thunder should be considered a signal to seek shelter immediately. The best way to remain safe from lightning strikes is go inside a sturdy building. A shed or flimsy structure will not protect you from lightning. Once indoors, do not talk on the telephone (cordless and cellular phones are safe); take a bath or shower; or handle electrical appliances, computers, or plumbing fixtures until the storm has passed. It's safest to unplug all electrical appliances except a radio, so you can stay apprised of severe weather alerts.

If you are not near a building when lightning threatens, the next safest option is to get into your car (as long as it's not a convertible!) and keep the windows rolled up. Do not touch the metal sides of the car.

If you are outdoors, far from buildings and vehicles, go to the lowest spot in the vicinity and crouch down. Keep away from trees, fences, and poles. If you are in the woods, stay away from the tallest trees.

If you are outdoors and feel your skin tingle, feel your hair stand on end, or hear clicking sounds, lightning may be about to strike. In that case, the safest position is to crouch down on the balls of your feet. Place your hands on your knees and your head down between your hands. If possible, pick up one foot and balance on the other. Do not lie down—the idea is to remain low while minimizing your body contact with the ground.

Once you have reached safest possible spot, remain there until the storm has passed. If lightning strikes nearby, it does not mean the danger is over. Lightning commonly strikes the same spot more than once during a storm. In fact, the top of the Empire State Building was struck fifteen times during a fifteen-minute span!

The worst place to be in a thunderstorm is in the water. If you are boating or swimming, hurry back to land and seek shelter. Other dangerous places to be when lightning strikes include: under a tree; on an athletic field or golf course; on a bicycle, tractor, or rider lawnmower; and on a mountain.

Thunderstorms

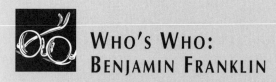

Who's Who: Benjamin Franklin

Benjamin Franklin was born in 1706 in Boston, Massachusetts, which at that time was a British colony. Franklin was the fifteenth child out of a total of seventeen. Because his family was poor, the young Franklin had only two years of formal education. Franklin made up for this deficiency by educating himself. He went on to become a scientist, diplomat, author, publisher, and inventor.

Franklin was a pioneer in the study of electricity. He first conducted experiments using a Leyden jar, which is a glass jar filled with water and plugged with a rubber stopper. It contains a metal rod inserted through the stopper, one end of which extends into the water. The other end of the rod is connected to a machine that generates an electric charge. Using the Leyden jar, Franklin studied the nature of static electricity in water and the glass that enclosed the water.

The crackling noise made by electricity in the Leyden jar reminded Franklin of the crackling of thunder. This observation led him to wonder if lightning was also a form of electrical discharge. Late in 1752, in Philadelphia, Pennsylvania, Franklin conducted his famous kite-flying experiment to test this hypothesis.

He fashioned a kite from two wooden sticks and a large silk handkerchief. He attached a metal key to the kite string, just above the point where he held the string, and set the kite flying during a

does. Lightning also is responsible for around 10,000 brushfires and forest fires annually, particularly in the western United States, western Canada, and Alaska. In addition, tens of millions of dollars in damage is caused to electrical utility equipment. The total property damage due to lightning in the United States alone is several hundred million dollars per year.

Thunderstorms

thunderstorm. The storm-generated electricity traveled down the rain-drenched string, to the key. When Franklin touched the key, he felt a shock.

Fortunately, Franklin had the foresight to run a wire from the key to the ground, so the electric charge would run into the ground. If he had not grounded his experiment in this way, the electrical discharge might have killed him. Franklin was also fortunate that lightning did not strike his kite directly. If that had happened, the grounding wire would probably have not protected him from a lethal dose of electric shock.

Three years prior to his famous experiment, Franklin had invented **lightning rod**s as a way to protect tall structures from lightning strikes. A lightning rod is a metal pole that is attached to the tallest point of a building and connected, by an insulated conducting cable, to a metal rod buried in the ground. Franklin's invention caught on quickly. Most tall structures, to this day, are topped with lightning rods.

Franklin's weather observations went far beyond the topic of lightning. In 1743, Franklin was the first to conclude that a local storm was not an isolated event, but rather was due to the large-scale circulation of **air mass**es. This discovery was made 175 years before meteorologists in Scandinavia discovered that rotating **front**s produce large, organized storm systems.

Franklin noticed that a storm had followed a path from Philadelphia to Boston—that is, from the southwest to the northeast. During the storm, however, the surface winds were blowing from the northeast. Franklin concluded that since the local storm had arrived from a direction counter to that of the local winds, it must not be local in nature, but part of a larger storm system.

Lightning is most often produced by **cumulonimbus** clouds during the **mature stage** of a thunderstorm. However, it can also arise from other clouds, including: **cumulus** clouds; **stratus** clouds; clouds produced by volcanic eruptions; or even billowing clouds of sand produced during sandstorms.

Lightning, for all of its ill-effects, is not without its benefits. First, it makes possible the conversion of normal oxygen to ozone. Ozone in the atmosphere protects plants and animals from harmful ul-

Thunderstorms

traviolet radiation. Second, lightning breaks down oxygen and nitrogen in the air, producing ammonia and nitrogen oxides. These chemicals react with rainwater to form nitrogen compounds, which are natural fertilizers. Over 100 million tons of nitrogen compounds fall to the ground each year.

And if it were not for lightning, we might not be here at all. Many scientists believe that lightning initiated the series of chemical reactions in the oceans that led to the formation of life on Earth.

HOW LIGHTNING IS PRODUCED. In order for lightning to occur, there must be two objects, or regions, carrying different electrical charges. If an object gains electrons, it is said to be "negatively charged." In contrast, if an object loses electrons it is said to be "positively charged." An object with any electrical charge, positive or negative, is **ionized.**

Lightning is a surge of electrons between ionized areas, each with different electrical charges. The majority of lightning occurs within a sin-

Figure 24: Charge distribution leading up to a lightning strike.

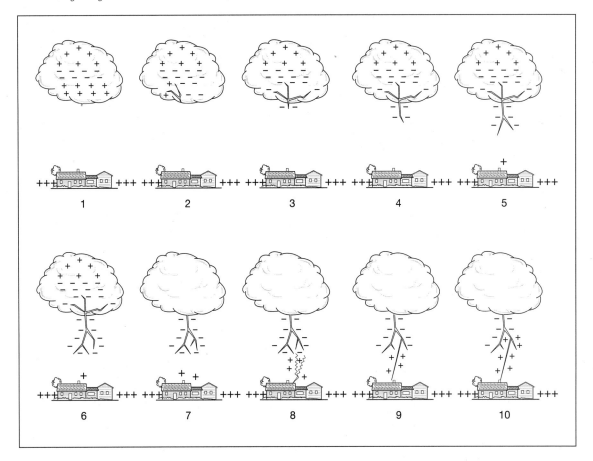

gle cloud. Most of the rest occurs between a cloud and the ground. Cloud-to-cloud or cloud-to-air lightning occurs infrequently.

Under normal conditions, such as on a clear day, the ground is negatively charged while the upper air is positively charged. When a cumulonimbus cloud forms, the charge distribution changes. The ground beneath the developing cloud becomes positively charged. A narrow region at the base of the cloud, as well as the upper portion of the cloud, also become positively charged. However, in the lower portion of the cloud, just above the base, there exists a negatively charged, saucer-shaped region. This region is about 1,000 feet (300 meters) thick and several miles in diameter.

Under most conditions, air acts as an **insulator,** meaning electricity does not readily flow through it. During the mature stage of a thunderstorm, however, the electrical charge differential between the cloud and the ground becomes so great that the resistance of air breaks down. Specifically, air loses its insulating properties when the electric field grows stronger than 915,000 volts per foot (about 3 million volts per meter). Then, electricity, which appears as lightning, surges between differently charged regions in order to neutralize the opposing charges.

Meteorologists are still not sure of the reason behind the distribution of charges in cumulonimbus clouds. One theory is based on the movement of hail or **graupel** throughout the cloud. As we learned in the chapter on "Precipitation," both hail and graupel grow within a cloud as **supercooled water** droplets and ice crystals freeze onto them. For simplicity in the following explanation, we will refer to hail and graupel collectively as "hail."

The reason that hail is believed to influence the distribution of charges within a cloud has to do with differences in temperature between hail and ice crystals. This temperature difference arises in the following way: **latent heat** is released when water droplets within a cloud freeze to a **hailstone.** This change makes the surface of a hailstone warmer than the ice crystals it encounters in the upper portion of the cloud. Since electrons flow away from a colder object and toward a warmer object, hail takes electrons away from the ice crystals. Specifically, when hail strikes ice crystals at temperatures below 5°F (-15°C), the hail becomes negatively charged while the ice crystals become positively charged.

The ice crystals, which are smaller than hail, get carried to the top of the cloud by **updraft**s. This may account for the positive charge in the upper region of the cloud. The hail, which is heavier, collects in the

Thunderstorms

WEATHER REPORT: THE COLOR OF LIGHTNING

Lightning takes on a range of colors, depending on atmospheric conditions.
- Blue lightning within a cloud indicates the presence of hail.
- Red lightning within a cloud indicates the presence of rain.
- Yellow or orange lightning indicates a large concentration of dust in the air.
- White lightning is a sign of low humidity in the air. Since forest fires break out when the air is relatively dry, white lightning is the most likely type to ignite forest fires.

lower region of the cloud, possibly accounting for the negatively charged area there. The negatively charged area near the base of the cloud repels electrons at the surface. This creates a positively charged area directly beneath the cloud, which moves along with the cloud.

CLOUD-TO-GROUND LIGHTNING. Lightning that travels between a cloud and the ground accounts for only 20 percent of all lightning, yet this type of lightning has been studied more extensively than any other. **Cloud-to-ground lightning** is considered the most important by researchers since it is the only type that endangers people and objects on the ground. Here we will trace the sequence of events that occurs during cloud-to-ground lightning.

Although lightning lasts only for two-tenths of a second and appears as a mere flash of light, it is quite a complex process. It begins when an invisible stream of electrons, called a **stepped leader,** surges from the negatively charged region of the cloud, down through the base. It is called a "stepped leader" because it travels in a stepwise fashion down toward the ground. Each portion of the stepped leader covers about 60 to 300 feet (20 to 100 meters) in less than a millionth of a second. Then it stops for about 50 millionths of a second before starting off in a new direction. The stepped leader creates a branching pattern, ionizing a path through the air as it goes.

When the stepped leader reaches a point about 300 feet (100 meters) above the ground, lightning surges up from the ground to meet it. The ground-based lightning is called a **return stroke.** It typically comes from a tall, pointed object, such as an antenna or flagpole, since the tallest objects in a region have the greatest positive charge.

Once the return stroke contacts the stepped leader, it completes the **ionized channel** between the cloud and the ground. A large concentration of electrons are discharged to the ground through this channel. Then positive ions from the ground shoot back up to the cloud. The upsurge of positive ions generates the bright flash commonly considered "lightning."

The return stroke is 2 to 7 inches (5 to 8 centimeters) in diameter. It travels at nearly one-third the speed of light and takes a mere one ten-thousandth of a second to reach the cloud. This flow of positive ions partially neutralizes the charge difference between the cloud and the ground.

In the approximately one-tenth of second that follows, several (usually two to four but sometimes as many as twenty) more lightning strokes may occur along the ionized channel. These strokes, which serve to discharge the remaining buildup of electrons near the base of the cloud, are initiated by surges from the base of the cloud called **dart leader**s.

Dart leaders, like stepped leaders, are intercepted by return strokes when they get closer than 300 feet (100 meters) to the ground. These return strokes occur about 50 thousandths of a second apart. They are individually indistinguishable and appear as a flickering light in the wake of the initial return stroke. The dart leaders cease when the charge differential between the cloud and the ground has been neutralized.

The most dangerous type of lightning is cloud-to-ground lightning.

Thunderstorms

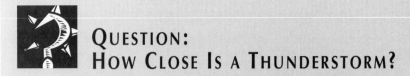

QUESTION: HOW CLOSE IS A THUNDERSTORM?

While we experience the visual component of lightning at almost the exact instant it occurs, we don't experience its audio component, thunder, until a short time later. The reason for this delay is that lightning travels at the speed of light (186,282 miles per second or 299,914 kilometers per second), which is about one million times faster than the speed of sound (1,100 feet per second or 330 meters per second).

One way to tell how close you are to a thunderstorm is to determine the time lapse between the lightning and the thunder. The rule to remember is that it takes thunder about five seconds to cover one mile (three seconds for one kilometer). Therefore, if you hear thunder seven seconds after you see the flash of lightning, the thunderstorm is 1.4 miles (2.3 kilometers) away (7 divided by 5 equals 1.4).

You can tell, in general terms, whether a thunderstorm is near or far, by the quality of the thunder. If it sounds like a sharp crack or clap, the storm is close. If that sound is immediately followed by a loud bang, the storm is *very* near—closer than 330 feet (100 meters).

The thunder from distant storms produces a rumbling sound. One reason for this effect is that the sound waves are bouncing off hills or buildings before reaching you. Another reason is that you first hear the sound from the part of the lightning near the ground, which is closer to you, after which you hear the sound from the upper part of the lightning, which is farther away.

In less than 10 percent of all cases of cloud-to-ground lightning, a positively charged stepped leader surges from the upper portion of the cloud. It travels downward to a negatively charged area on the ground. These powerful discharges occur most commonly during winter storms and produce a flash of light similar to a return stroke.

Ground-to-cloud lightning occurs even less frequently. This form of lightning begins with the ascent of a stepped leader, usually positively charged, from the ground. As the stepped leader approaches the cloud above, it triggers the release of a return stroke from the cloud. Ground-to-cloud lightning is most often initiated from very tall points on the surface, such as mountaintops or the tops of towers or antennae.

OTHER TYPES OF LIGHTNING. Cloud-to-cloud lightning is the most common form of lightning. It occurs either within a single cloud or between two clouds. In the former case, the lightning runs between the negatively charged lower portion of the cloud and the positively charged upper portion of the cloud. This type of lightning illuminates and provides a brilliant view of a cumulonimbus cloud.

Lightning that runs between two clouds occurs less frequently than it does within a single cloud. Inter-cloud lightning represents a discharge of electrons from the lower portion of one cloud to the upper portion of an adjacent cloud.

Cloud-to-air lightning is the flow of electricity between areas of a cloud and the surrounding air which have opposite charges. This form of lightning is relatively weak and often occurs at the same time as a cloud-to-ground stroke. Usually, this lightning travels a path between an area of negative charge in the surrounding air and the positively charged top of the cloud. Because cloud-to-air lightning occurs at great heights, it is almost always too distant to have an audible thunder component.

Both cloud-to-cloud lightning and cloud-to-air lightning are often referred to as **sheet lightning.** Sheet lightning illuminates a cloud or a portion of a cloud. The cloud blocks the distinct pattern of the lightning from view, so the lightning appears as a bright sheet.

Ball lightning is the rarest and most mysterious form of lightning. It has never been photographed but has been witnessed by numerous individuals throughout history. It is reported to look like a dimly-to-brightly lit sphere, ranging from .4 to 40 inches (1 to 100 centimeter) in diameter. It lasts between one and five seconds and either hangs in the air or floats horizontally at a rate of about 10 feet (3 meters) per second. It either dissipates silently or with a bang.

The cause of ball lightning is unknown, but many theories have been proposed. One recent theory suggests that it is an "electromagnetic knot" created by linked lines of magnetic force that form in the wake of an ordinary cloud-to-ground lightning strike. Some scientists suggest that ball lightning does not exist, but is merely an optical illusion experienced by an individual who has just witnessed a stroke of lightning.

Thunderstorms

WEATHER REPORT: LIGHTNING RODS

Lightning rods are metal poles used to protect buildings from lightning strikes. A lightning rod is attached to the tallest point of a structure and connected, by an insulated conducting cable, to a metal rod buried in the ground. The principle behind lightning rods is that since lightning generally strikes the tallest target, it will strike the rod and pass into the ground, sparing the building.

A lightning rod provides protection to a cone-shaped area around and beneath it. The tip of the cone coincides with the top of the rod. The radius of the base of the cone is equal to the height of the rod. Thus, the taller the lightning rod, the greater the area it protects.

When a return stroke originates from two different places on the ground at once, it creates two separate ionized channels. This type of lightning is called **forked lightning.**

Ribbon lightning is lightning that appears to sway from a cloud. It is produced when the wind blows the ionized channel so that its position shifts between return strokes.

A flash of lightning that resembles a string of beads is called **bead lightning.** This type of lightning may be the result of a fragmenting ionized channel. An alternative explanation is that part of the lightning stroke is obscured by clouds or falling rain.

Silent lightning from a distant storm, generally more than about 10 miles (16 kilometers) away is called **heat lightning.** This lightning is not accompanied by thunder since it is too far away for the sound to reach the observer. It often occurs on hot summer nights when the sky above is clear. Heat lightning sometimes appears orange, due to the scattering of light by dust particles in the air.

Downbursts

A **downburst** is an extremely strong, localized **downdraft** beneath a **thunderstorm.** It blasts down from a **thunderstorm cloud** like water pouring out of fully opened tap. When this vertical wind hits the ground it spreads horizontally. It then travels along the ground, destroying objects in its path.

Downbursts may or may not be accompanied by rain. Dry downbursts generally occur beneath **virga,** streamers of rain that evaporate in mid-air. Downbursts are capable of knocking down trees, damaging buildings, and leveling crops, as well as kicking up dust and debris into a cloud that tumbles along the ground. In many cases, damage caused by downbursts has been mistakenly attributed to **tornado**es.

Downbursts are divided into two categories—**macroburst**s and **microburst**s—depending on their size. Downbursts may result in long-lived, tornado-force, straight-line winds called **derecho**s. For a discussion of derechos, turn to the chapter entitled "Local Winds."

MACROBURSTS. A macroburst is a downburst that creates a path of destruction on the surface greater than 2 miles (4 kilometers) wide. The winds of a macroburst travel at around 130 mph (210 kph) and last for up to thirty minutes. A macroburst (as well as a microburst) may either follow in the wake of, or its leading edge can develop into, a **gust front.**

MICROBURSTS. A microburst is smaller, yet potentially more dangerous, than a macroburst. The path of destruction created by a microburst is between several hundred yards and 2 miles (4 kilometers) wide. Its winds, which only last about ten minutes, may exceed 170 mph (270 kph). Like macrobursts, microbursts may evolve into gust fronts.

Microbursts receive more attention than macrobursts because of the hazard they pose to airplanes during take-off or landing. Microbursts are accompanied by abrupt changes in the speed or direction of wind at various heights, known as **wind shear.** Wind shear is something that every pilot seeks to avoid, since it can spell disaster for aircraft. And due to their small size, microbursts frequently elude detection by airport **radar.**

As a plane enters a microburst, it first experiences a strong headwind that sends it upward. Soon thereafter, the plane experiences a strong tailwind which forces it downward. In the thirty-year period from 1964 to 1994, about thirty planes have crashed as a result of microbursts. After pilot error, microbursts are the second-leading cause of airplane crashes.

Microbursts occur with an alarming frequency. In a 1978 study, conducted by T. Theodore Fujita (see box, page 238), fifty microbursts

Thunderstorms

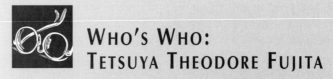

Who's Who: Tetsuya Theodore Fujita

Although Tetsuya Theodore Fujita is most famous for developing the scale of tornado intensity which bears his name (see page 252), his primary area of research has been downbursts. Fujita was the first to identify these destructive downdrafts of wind. His research on downbursts has been particularly relevant to aviation safety, since microbursts, the smallest and most intense form of downbursts, pose an extreme hazard to aircraft.

Fujita was born in 1920 in Kitakyushu City, Japan. He graduated from Meiji College of Technology with the equivalent of a bachelor's degree in mechanical engineering in 1943. Soon thereafter, Fujita was hired on at the college as an assistant professor of physics.

In 1945, the Japanese cities of Hiroshima and Nagasaki were devastated by atom bombs dropped by United States airplanes. Three weeks later, Fujita was part of a research team sent to those cities to survey the damage. Fujita noticed that the destruction was in the shape of a starburst. The hub was located directly beneath the bomb and spokes radiated outward. Fujita also calculated the height from which the bombs must have been dropped to create such a pattern. These findings became relevant in Fujita's later work on downbursts.

were detected over forty-two days in Chicago's western suburbs. Another study was conducted near Denver's Stapleton International Airport over an eighty-six-day period in the summer of 1982. A total of 186 microbursts were detected.

Flash Floods

A **flash flood** is a sudden, intense, localized flood caused by persistent, torrential rainfall. In the 1970s, flash floods replaced **lightning** as

> In 1949, Fujita moved to Tokyo to pursue his doctorate in atmospheric science at Tokyo University. In 1953, at the invitation of professor and thunderstorm-specialist Horace R. Byers, Fujita moved to the United States to join the faculty of the University of Chicago.
>
> Fujita's main topic of research soon became the destructive potential of storm-related winds, particularly tornadoes. Based on his surveys of tornado damage, Fujita created the **Fujita Intensity Scale** for tornadoes in the late 1960s. Fujita's scale consists six categories of tornado intensity, based on the ground damage created by the tornado. His scale provided the first objective, uniform way of assessing tornado strength.
>
> In April 1974, Fujita took a plane ride to survey the damage caused by the Super Tornado Outbreak. Flying over West Virginia, he noticed the same starburst-pattern of destruction he had seen in Japan. Fujita proposed that in that area, the damage had been created not by a tornado, but by powerful downdrafts produced by thunderstorms. He then coined the term "downbursts."
>
> At first, Fujita's findings were met with skepticism by his fellow meteorologists. It was commonly accepted at the time that thunderstorms produce downdrafts of air, but it was believed that downdrafts weaken significantly before reaching the ground. Fujita conducted a research project to put his hypothesis to the test. Over a forty-two day period in the spring and summer of 1978, Fujita and his team of researchers detected fifty microbursts in Chicago's western suburbs.
>
> Fujita retired from teaching in 1991. He still conducts research on destructive winds and ways to improve air safety.

the number-one weather-related killer in the United States. Each year throughout the 1980s, flash floods killed an average of 110 people and were responsible for an average of $3 billion in property damage. In the first half of the 1990s the number of deaths due to flash floods rose to an annual average of 140.

Flash floods are caused by warm-weather **thunderstorm**s that are either slow-moving or stationary. The reason for their lack of movement is that the **winds aloft** are nearly calm. These storms unleash huge quantities

Thunderstorms

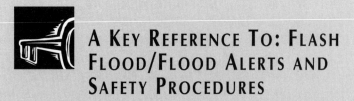

A Key Reference To: Flash Flood/Flood Alerts and Safety Procedures

A flash flood **watch** means that heavy rains may cause flash flooding within the designated area. This is the time to prepare to evacuate to higher ground.

A flash flood **warning** means that a **flash flood** has been reported and is imminent for the designated area. In some areas, a warning signal sounds when waters reach critical levels. When a warning is issued, it's imperative to move to safe ground immediately. You may only have seconds or minutes before waters become dangerously high.

If you live in an area that's prone to flash flooding or flooding, it's wise to take the following precautions:
- Learn the evacuation routes to higher ground.
- Keep your car's gas tank filled.
- Put together a supply kit for your home containing first aid materials, a battery-powered flashlight, a battery-powered radio, extra batteries, rubber boots, rubber gloves, non-perishable food, a non-electric can opener, and bottled water.
- Install check valves in your home's sewer traps to prevent flood water from backing up and coming in through your drains.

of rain over one location. Thunderstorms that move more quickly, in contrast, spread their rain across larger areas and don't produce flash floods.

A flash flood is initiated when the amount of rainfall exceeds the capacity of the ground to absorb it. In that case, the rainwater runs along the surface, rather than sinking into it. It flows to the lowest point, which is generally a river, stream, or storm sewer. If the quantity of water is greater than the capacity of the drainage channel, within a few minutes the channel overflows and a flash flood occurs.

As the banks overflow, water rushes forward at speeds of up to 20 mph (35 kph). It takes the form of a sediment-laden wall that can surge as high as 30 feet (10 meters). The floodwaters are capable of dislodging

Thunderstorms

> Once a flood watch has been issued, fill your bathtub and large containers with water for drinking and cooking in case your local water source becomes contaminated.
>
> If you encounter a flash flood, the most important thing to remember is to avoid the water. If you are driving, turn around and go the other way. If you are on foot, climb to higher ground. Do not try to walk, swim, or drive through flood water. The water is moving rapidly and carries dangerous debris. You can be knocked over by as little as 6 inches of rushing water. Deeper water may have an undertow that drags you beneath the surface.
>
> If you are in a car during a storm, watch for flooding where the road dips down, under bridges, and in low areas. These are the places that fill with water first. If you see a flooded area, turn the vehicle around and go the other way. Be especially careful at night, when the beginnings of flash floods are harder to recognize.
>
> If your car stalls or becomes blocked, abandon the car and seek higher ground. A car will float away in just 2 feet of water. Then it is at the mercy of the current, which may carry it into deeper water or overturn it in a ditch. Sixty percent of people who die in flash floods are either in a car or are attempting to leave a car that has been stranded in high water.

large objects weighing several tons, such as boulders, cars, and even train engines. These objects may be carried several miles downstream. Flash floods are also known to tear out bridges in their path.

Flash floods are common in mountainous areas. Rain runs down mountainsides and becomes concentrated in canyons and valleys. Arid and urban regions are also prone to flash floods. The reason for this fact is that very little rain can seep into parched ground or concrete. Most cities have storm sewer systems that direct water underground to nearby rivers. If the storm sewers become clogged or overwhelmed by the volume of water, the streets quickly become flooded.

Flash flooding may also result from rains associated with a **hurricane,** a break in a dam or levee, or the springtime melting of large quantities of snow and ice.

Thunderstorms

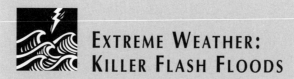

Extreme Weather: Killer Flash Floods

- The deadliest flash flood in the history of the United States occurred in Johnstown, Pennsylvania, in May 1889. This flood, which killed 2,209 people, was due to the collapse of a 72-foot-high, 930-foot-long dam. Johnstown was hit with a 23-foot-tall wall of water.
- One of the most destructive flash floods in recent history occurred in Colorado's Big Thompson Canyon, 50 miles (80 kilometers) west of Denver, in July 1976. Ten to 12 inches (25 to 30 centimeters) of rain, over half the yearly average rainfall for that location, caused the overflow of the river running through the narrow canyon floor. A wall of water 20 feet (6 meters) high rushed forward, killing at least 139 people. More than 400 houses were destroyed, roads were washed away, and 197 vehicles were lost in the flood. About 1,000 people were rescued by helicopter.
- In June 1990, a flash flood occurred in Shadyside, Ohio, when 4 inches (10 centimeters) of rain fell in less than two hours. The resultant 30-foot (9-meter) tall wall of water killed 26 people and caused around $7 million in damages.
- Cheyenne, Wyoming, was beset by a flash flood in August 1985, after 6 inches (15 centimeters) of rain fell in three hours. The streets were filled with more than 6 feet (less than 2 meters) of water. The flood caused the deaths of 12 people and more than $65 million in damage.

Tornadoes

In this chapter we will look closely at the most violent offspring of **severe thunderstorm**s: **tornado**es. In the early part of this century, tornadoes killed an average of 200 people each year in the United States. Due to improvements in forecasting and emergency preparedness in recent years, this number has been drastically reduced. Over the last thirty years, there have been an average of only 82 deaths per year from tornadoes. The number of fatalities due to tornadoes in the United States is now less than the number due to two other thunderstorm-related phenomena: **flash flood**s and **lightning.**

Tornadoes accompany only 1 percent of all thunderstorms. An average of 800 to 900 tornadoes occur yearly in the United States, although the numbers vary widely from year to year. They are short-lived and often strike sparsely populated areas.

Characteristics of Tornadoes

A tornado is a rapidly spinning column of air that extends from a thunderstorm cloud to the ground. The tornado rotates along a vertical axis of extremely low pressure called a **vortex.** Tornadoes are sometimes called "twisters" or "cyclones."

The use of the word "cyclone" to describe a tornado, however, can be confusing. A **cyclone** is actually any storm system in which winds spiral around a low-pressure area. A tornado is merely one type of cyclone. Other types of cyclones include **hurricane**s and large-scale storms that occur in the **middle latitudes.**

Tornadoes

A tornado starts out as a **funnel cloud** that pokes through the base of a **thunderstorm cloud** and hangs in the air. Only when the funnel cloud touches the ground is it called a tornado.

Tornadoes vary widely in appearance. Probably the most familiar-looking type is a cylindrical-shaped funnel. Tornadoes may also resemble upside-down bells or long, thin rope-like pendants. Others take the shape of an elephant's trunk.

Tornadoes may be white, gray, brown, or black. If the air sucked into the vortex cools to its **dew point** while traveling up to the cloud, then a cloud forms within the tornado, giving it a white or gray appearance. Debris and dust that gets picked up from the ground and spun up, into the vortex, sometimes colors the tornado dark-brown or black. Occasionally, a tornado will even take on a reddish color because of red dirt or clay along the tornado's path.

A large tornado on the ground near Union City, Oklahoma.

Slightly less than 70 percent of all tornadoes are considered "weak tornadoes," with spinning winds of 75 to 110 mph (120 to 180 kph). Those tornadoes create a damage path that is 30 to 200 feet (10 to 60 meters) wide by 5 miles (8 kilometers) long, on average. They make contact with the ground for less than ten minutes and account for less than 5 percent of all tornado deaths.

Nearly 30 percent of all tornadoes are considered "strong tornadoes," with spinning winds of 110 to 200 mph (180 to 320 kph). Strong tornadoes create a path of destruction that is up to 1 mile wide and several miles long. They make contact with the ground for around twenty minutes and account for almost 30 percent of tornado deaths.

Only 2 percent of all tornadoes are considered "violent tornadoes." Their spinning winds travel at speeds surpassing 250 mph (400 kph), gusting up to 320 mph (500 kph). They create a path of destruction that is greater than 1 mile wide and over 100 miles (160 kilometers) long. A tornado of this intensity can last more than two hours. Violent tornadoes, despite the infrequency with which they occur, account for 70 percent of all tornado deaths.

Tornadoes travel along the ground at an average speed of 34 mph (55 kph). On one extreme there are tornadoes that barely inch along, while on the other extreme there are tornadoes that race through at 150 mph (240 kph).

Tornadoes are spawned by, and travel with, severe thunderstorms. These thunderstorms are usually of the **supercell** variety. The destructive force of a tornado is generated by the extreme difference in **air pressure** between the center of a tornado and its outer edges. The difference in air pressure within a tornado is 10 percent, which is similar to the difference in air pressure between a location at sea level and a location .6 mile (1 kilometer) above sea level.

Most tornadoes are spawned from thunderstorms at the leading edge of a **cold front.** The thunderstorm and tornado follow the general path of the cold front, which is from southwest to northeast. It is difficult to predict exactly where a tornado will strike, however, because tornadoes seldom move in a straight line. They frequently change direction and sometimes even move in circles or figure-eight patterns. The ability of forecasters to predict a tornado has improved in recent years, with the advent of **Doppler radar.** (See "Forecasting," page 414.)

A group of tornadoes that develops from a single thunderstorm is called a **tornado family.** These are quite rare. When they do form, it is

Tornadoes

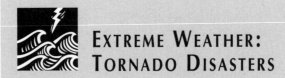

EXTREME WEATHER: TORNADO DISASTERS

- The world's deadliest tornado outbreak ever recorded, known as the "Tri-State Tornado Outbreak," was on March 18, 1925. Seven tornadoes (according to some sources, eight) traveled through parts of Missouri, Illinois, and Indiana. The tornadoes killed 695 people and injured 2,027 along their combined 437-mile-long (703-kilometer-long) path. In one town, Murphysboro, Illinois, 234 people were killed. Two other towns, Gorham, Illinois, and Griffin, Indiana, were totally destroyed.
- Another famous outbreak, known as the "Palm Sunday Outbreak," came on April 11, 1965. Forty tornadoes ravaged parts of Wisconsin, Iowa, Illinois, Indiana, Michigan, and Ohio. In that outbreak, 256 people lost their lives and over 3,000 were injured. Property damage amounted to around $300 million.
- The most destructive tornado outbreak in the history of the United States, referred to as the "Super Tornado Outbreak," occurred on April 3 and 4, 1974. Over a period of 16 hours, 148 tornadoes twisted their way through parts of 13 states, from Mississippi to Michigan. At least 6 of the tornadoes were "violent," with winds exceeding 260 mph (420 kph). The tornadoes caused a total of 315 deaths, over 6,100 injuries, and $600 million in property damage. They created a combined path of destruction that was 2,598 miles (4,180 kilometers) long.
- A tornado outbreak struck the Mararipur district, 80 miles (130 kilometer) outside of Bangladesh's capital city, Dhaka, on April 1-2, 1977. The tornado outbreak left 500 people dead, over 6,000 injured, and hundreds of thousands homeless.
- On March 28, 1984, an outbreak of 36 tornadoes killed 59 people in North Carolina and South Carolina. Of those killed, 37 percent were inside mobile homes. More than 1,200 people were injured and damage costs exceeded $200 million.
- The deadliest single twister ever recorded was deadlier than any tornado outbreak. It hit Dhaka, Bangladesh, on April 26, 1989. The tornado killed at least 1,109 people, injured about 15,000, and left about 100,000 homeless.

usually along a **squall line.** The emergence of a tornado family is referred to as a **tornado outbreak.** Tornado outbreaks are responsible for the greatest amount of tornado-related damage. A single outbreak can result in over 100 deaths. The tornadoes in an outbreak touch down at random locations and create a patchwork pattern of destruction.

FORMATION OF TORNADOES

The processes that lead to tornadoes remain somewhat mysterious to meteorologists. Because the emergence and paths of tornadoes are unpredictable, and because tornadoes are usually short-lived, they make difficult subjects to study. And it's nearly impossible to learn about a tornado by looking directly into the funnel!

Until recently, there has been no way to determine the direction and speed of winds within a tornado. Conventional weather instruments, such

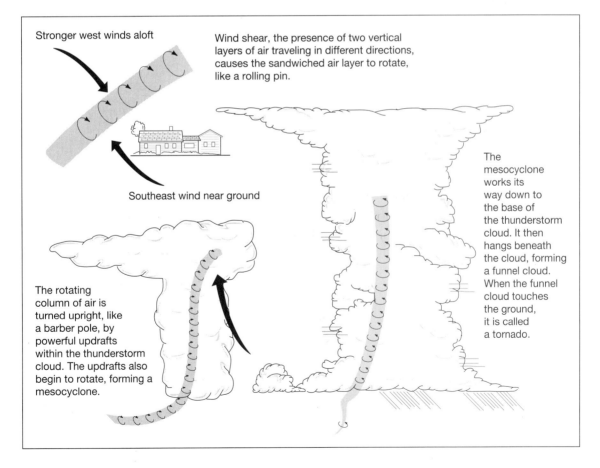

Figure 25:
Formation of a tornado.

Tornadoes

as **anemometer**s, cannot withstand the extreme conditions of a tornado. Only since the advent of Doppler radar in the mid-1970s have scientists been able to measure a tornado's winds. Even then it's a difficult task, since Doppler units must be positioned in the right place and at the right time in order to obtain measurements.

In recent years, using the Doppler measurements and computer modeling systems, meteorologists have come up with a likely scenario of how tornadoes are formed. As we have already stated, tornadoes originate within severe thunderstorms, most often supercell storms. Here we will pick up where we left off in the section on supercell thunderstorms (see "Thunderstorms," page 223), with the formation of a **mesocyclone**.

A mesocyclone is a region of rotating **updraft**s within a mature thunderstorm. The mesocyclone begins in the middle portion of the thunderstorm and works its way downward, toward the base. As air from below is drawn into the mesocyclone, the mesocyclone becomes longer

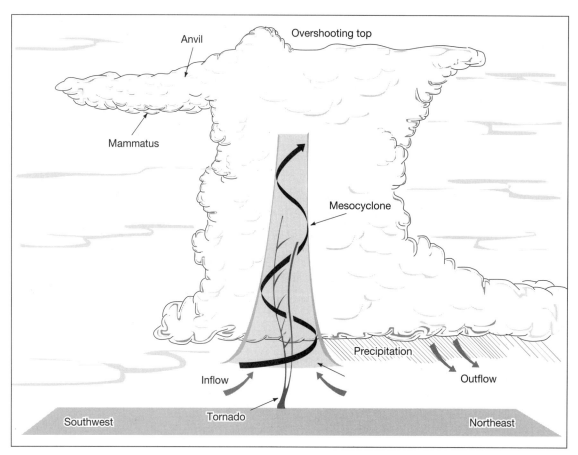

Figure 26: Tornado position within a thunderstorm.

248 THE COMPLETE WEATHER RESOURCE

and skinnier. The diameter of the mesocyclone shrinks to about 3 to 6 miles (5 to 10 kilometers). At the same time, due to the **conservation of angular momentum** (see "What Is Weather?" on page 42), the mesocyclone spins faster and faster. When this spinning column protrudes through the cloud base it is no longer considered a mesocyclone, but a **tornado cyclone.**

The next step is the formation of a funnel cloud. A funnel cloud is a cone-shaped tornado cyclone that hangs below the base of the cloud. The funnel cloud grows as updrafts of air rush into the zone of extremely low pressure in the vortex of the tornado cyclone. When the incoming air rises, it cools. If the air is moist enough, **condensation** occurs and forms a cloud, making the funnel cloud visible. The funnel cloud may continue to build downward. If it reaches the ground, it will be called a tornado.

Most strong and violent tornadoes develop in the way just described within the mesocyclone, which is located in the right rear area of a severe thunderstorm. However, relatively weak, short-lived tornadoes may also form along the thunderstorm's **gust front.** Unlike strong tornadoes, which form in the thunderstorm's updrafts, weak tornadoes are created in the **downdraft**s.

LIFE CYCLES OF TORNADOES

Similar to the process by which thunderstorms form, tornadoes evolve through a number of stages. The first stage is called the **dust-whirl stage.** This stage is marked by the formation of a short funnel cloud protruding downward from the base of the thunderstorm cloud. The funnel cloud causes the swirling around of debris on the ground beneath it. Surface winds in this stage are rarely strong enough to cause any damage.

The second stage is called the **organizing stage.** During this stage, the funnel cloud extends further downward and increases in strength.

A tornado is at its most destructive in the **mature stage.** The funnel reaches all the way to the ground in this stage and remains in contact with the ground until it dissipates. During this stage, the tornado attains its greatest width and is nearly vertical. This stage lasts only fifteen minutes, on average.

During the mature stage, some tornadoes become **multi-vortex tornado**es. In these tornadoes, the vortex divides into several smaller vortices called **suction vortices.** Suction vortices are responsible for the

Tornadoes

strongest surface winds. They continually form and dissipate as the powerful tornado moves along.

As a tornado enters the **shrinking stage,** its funnel narrows. Friction with the ground causes the funnel to tilt. In this stage, which lasts an average of seven to ten minutes, the tornado creates a narrower path of destruction than it did during the mature stage.

The last stage is called the **decay stage.** In this stage, the funnel narrows further until it is shaped like a rope. It usually twists and turns several times before fragmenting and dissipating.

Not all tornadoes proceed through these five stages. Minor tornadoes may dissipate after the organizing stage, or proceed directly from the organizing stage to the decay stage.

Where and When Tornadoes Occur

Tornadoes occur wherever severe thunderstorms occur, primarily in temperate, middle latitude locations. About 75 percent of the world's tornadoes strike within the continental United States. The United States also receives more severe tornadoes, on average, than anywhere else in the world. Tornadoes in the United States are generated by the interaction of cold, dry air heading south from Canada and warm, moist air heading north from the Gulf of Mexico.

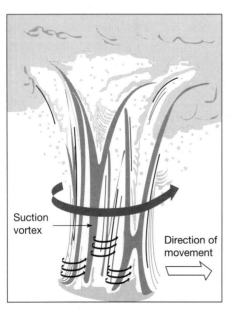

Figure 27: The suction vortices of a multi-vortex tornado.

Tornadoes are also common in Australia, New Zealand, South Africa, Argentina, and much of middle Europe as far south as Italy and as far north as Great Britain. Occasional tornadoes have been noted as far north as Stockholm, Sweden, and Saint Petersburg, Russia. Tornadoes occur infrequently in Japan, eastern China, northern India, Pakistan, and Bangladesh. Very weak tornadoes sometimes form in the tropics.

In the United States, the state that experiences the greatest number of tornadoes is Texas. The region that experiences the highest concentration of tornadoes per-area, known as "tornado alley," includes north-central Texas, Oklahoma, Kansas, Nebraska, and South Dakota. Central Oklahoma, in particular, has the greatest number of tornadoes per acre. Tornadoes are also common throughout the Mississippi Valley and in the Midwest, east to Massachusetts. They seldom occur west of the Rocky Mountains.

Tornadoes

Tornadoes occur during every month of the year in the United States. They occur with the greatest frequency in the spring and early summer months, April through June, and with the lowest frequency in December and January. The seasonal maximum varies with different parts of the country. For instance, in the lower Great Plains states (Kansas, Oklahoma, and Texas) and from Iowa east to Ohio, the greatest number of tornadoes occur in April through June. In Nebraska and South Dakota, as well as in Pennsylvania, the seasonal maximum is May through August. In North Dakota, Michigan, and the New England states, tornadoes peak in June through August. In the Gulf Coast states—Louisiana, Mississippi, and Alabama—tornadoes peak in intensity in March through May, with a secondary peak in November.

Tornadoes strike at all times of day and night. The peak time of day for tornadoes, during which 40 percent of all tornadoes occur, is between 2 P.M. and 6 P.M. Tornadoes occur with the least frequency just before sunrise.

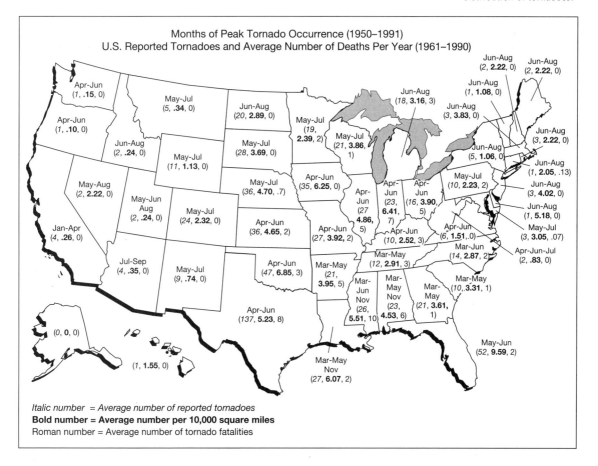

Figure 28: Geographic distribution of tornadoes.

Tornadoes

EFFECTS OF TORNADOES

Where a tornado touches the ground, it creates a path of near-total destruction. Primarily the high winds, and to a lesser extent the low-pressure, of a tornado cause walls to buckle and roofs to be lifted off and carried away. The winds can uproot trees and toss large, very heavy objects.

People, animals, and all matter of household items have been picked up by a tornado, only to be deposited several miles away. Entire buildings have been carried for hundreds of yards. And in 1931 a tornado picked up an 83-ton train car with 117 passengers on board and dumped it in a ditch 82 feet (25 meters) away.

There have been instances of frogs and toads being sucked up from a pond by a passing tornado, only to rain down on a community further along the storm's path. A motel sign in Broken Bow, Oklahoma, was carried for 30 miles (40 kilometers) by a tornado, before crashing down in Arkansas. And a canceled check was blown by a tornado for 305 miles (491 kilometers), from Great Bend, Kansas, to just outside of Palmyra, Nebraska.

MEASURING TORNADO INTENSITY: THE FUJITA SCALE

In the 1996 popular film *Twister,* a crew of storm chasers is hot on the trail of an F5 tornado, F5 being nature's most violent breed. What the "F5" refers to is the tornado's ranking according to the **Fujita Intensity Scale,** or the Fujita Scale for short.

Figure 29: The power of tornado winds.

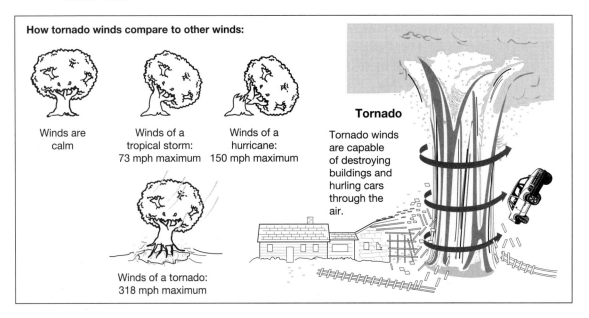

Weather Report: The Exploding Chicken Feathers

A well-known tornado-related oddity is the de-feathering of chickens. Reports of naked chickens abound after tornadoes strike rural areas. Professional and amateur meteorologists alike have been perplexed by this phenomena for years. The popular theory that the low-pressure of the tornado causes the feathers to "explode" off a chicken has been shown to be false. The fact is that a pressure drop great enough to cause feathers to explode off a chicken would also cause the whole chicken to explode.

A more likely explanation for de-feathering is that the frightened chicken induces what is called the "flight molt" response. This self-protective response, evoked by a chicken that is being threatened by a predator, causes the chicken's feathers to loosen. That way, when the predator chomps, it gets a mouthful of feathers instead of a mouthful of chicken.

During a tornado, the feathers loosened by the terrified chicken's flight molt response are merely blown away.

For more stories of tornado oddities, visit the Tornado Project Online at http://www.tornadoproject.com

---

The Fujita Scale divides tornadoes into six categories, F0 through F5, based on wind speed and damage created. The greatest significance of the Fujita scale is that it provides a universally accepted means of describing the intensity of tornadoes. Before the Fujita Scale was created, all tornadoes, from the weakest to the fiercest, were lumped together in official tallies. The Fujita scale also enables one to estimate the wind speed of a tornado by observing the damage that a tornado produces.

The Fujita Scale was developed in the late 1960s by T. Theodore Fujita, a meteorology professor at the University of Chicago (see "Thunderstorms," page 238). He was assisted by Allen Pearson, former director of the National Severe Storm Forecast Center. For this reason, the scale is sometimes referred to as the Fujita-Pearson Scale.

According to the Fujita Scale, F0 and F1 tornadoes are described as "weak"; F2 and F3 tornadoes are described as "strong"; and F4 and F5

Tornadoes

tornadoes are described as "violent." As mentioned earlier in this chapter, the "weak" category encompasses about 69 percent of all tornadoes, the "strong" category encompasses about 29 percent, and the "violent" category encompasses about 2 percent (only one or two tornadoes per year). It is the rare violent tornadoes, however, that account for the majority of tornado-related deaths and damage.

On the weakest end of the scale is an F0 tornado. An F0 tornado creates light damage on the order of broken branches and damaged billboards. It has winds less than 72 mph (115 kph). On the other extreme is an F5 tornado, which does incredible damage. An F5 tornado, with winds greater than 260 mph (420 kph), uproots trees, sends cars flying, and lifts houses off their foundations.

It is important to remember that the Fujita scale rates tornadoes according to intensity, not size. Many people believe that the largest tornadoes are the strongest and that the smallest tornadoes are the weakest. This is not always the case. A Fujita label can not be applied to an oncoming tornado. It can only be applied to a tornado once it has passed and the damage has been assessed.

WHEN TORNADOES APPROACH

If you see the sky turn a dark greenish color during a thunderstorm and/or large hail begins to fall, you may be in for a tornado.

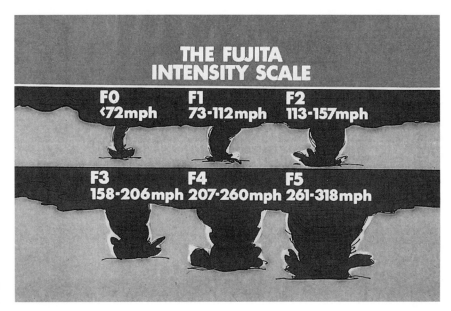

Figure 30: The Fujita Intensity Scale.

Tornadoes

As a tornado approaches, a portion of the base of a thunderstorm cloud will begin to rotate. This rotating area of cloud then lowers, forming a **wall cloud**. A wall cloud is a roughly circular cloud, 1 to 4 miles (1.5 to 6.5 kilometers) in diameter, that hangs beneath the thunderstorm cloud. The wall cloud can appear up to an hour before the tornado strikes.

The next sign is the appearance of a funnel at the base of the wall cloud. The funnel looks like a dimple bulging from the cloud and quickly lengthens. A funnel cloud will be visible only if moisture in the air rising within it condenses, forming a cloud. Another indication that a funnel cloud is present is the swirling of debris on the surface. Sometimes falling rain or darkness will obscure all signs of a funnel cloud.

Many tornadoes are said to be accompanied by a roaring sound, like several freight trains, loud enough to be heard miles away. Other tornadoes, however, travel quietly. A curious effect of tornadoes, reported by people who have been very near to their path, is a foul odor. Some have described it as an overpowering, sulfur-type of smell. There is no scientific explanation, at this point, for the source of such an odor.

WATERSPOUTS

Waterspouts are considered the first cousins of **tornado**es. In some cases, however, a waterspout *is* a tornado. A waterspout is a rapidly

Tornado damage: this business's owner kept his sense of humor.

Tornadoes

WHO'S WHO: HOWARD BLUESTEIN AND THE STORM CHASERS

The 1996 motion picture blockbuster *Twister* popularized the sport of **storm chasing,** zipping around in a vehicle in the hopes of sighting an elusive tornado. But real storm chasers, like University of Oklahoma professor Howard Bluestein, know that out there in the field, things are quite different from what was portrayed in the movie. The most important difference between the real world and the movie is that storm chasers do not place themselves in the path of the oncoming storm.

When he is out in the field, Bluestein plays it safe. "I think that we pretty much know where a tornado might form, and we're very careful not to get in its path," said Bluestein in an interview published in the April/May 1996 issue of *Weatherwise* magazine. "The greatest dangers that we have are dangers of driving under bad conditions. Roads that are wet. Narrow country roads.... Lightning is also a big scare. We try to stay inside the van if we see a lot of lightning."

Bluestein attended the Massachusetts Institute of Technology, where he majored in electrical engineering until his senior year. Then he switched to

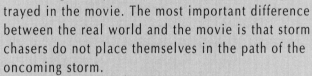

rotating column of air that forms over a large body of water. It extends from the base of a cloud to the surface of the water.

Contrary to the popular misconception, water is *not* drawn upward through the funnel, into the cloud. Rather, the moisture that exists within a waterspout is cloud droplets that have condensed from the rising air. At the point where the funnel comes in contact with the surface, the water sprays several feet upward.

There are two types of waterspouts. The first, called a **tornadic waterspout,** is merely a tornado that forms over land and travels over the

256 THE COMPLETE WEATHER RESOURCE

meteorology. He went on to complete a Ph.D. in tropical meteorology. Bluestein joined the faculty of the University of Oklahoma in 1976, where he has been ever since.

Bluestein has been leading teams of storm chasers, armed with cameras and portable Doppler radars, throughout the Oklahoma countryside since 1977. They are seeking answers to questions such as: Why do some thunderstorms spawn tornadoes while others do not? And how exactly do tornadoes form?

"You need a source of rotation and an updraft to spin it up. Everyone knows that," explained Bluestein in an interview published in the July 12, 1996, *Chronicle of Higher Education,* when asked about tornado formation. "What people don't agree on are the sources of rotation and the sequence of events that leads to the rotation, which produces the seeds of the tornado. We need to understand where the original vortex comes from."

Bluestein and his tornado-chasing colleagues have greatly advanced our understanding of tornadoes. Before tornado chasers began their quest two decades ago, all that was known about tornado formation was that tornadoes originated within supercell thunderstorms. The data collected by storm chasers has enabled forecasters to develop the computer models that more accurately, and farther in advance, predict when and where tornadoes will strike.

For more information on storm chasing, check out:
- *"Storm Chaser Homepage" on the Internet at: http://taiga.geog.niu.edu/chaser/*
- *Two video documentaries by Martin Lisius: "The Chasers of Tornado Alley" (1996) and "Chasing the Wind" (1991)*

water. Tornadic waterspouts are relatively rare and constitute the most intense form of waterspouts.

The vast majority of waterspouts form over the water and are called **fair-weather waterspout**s or non-tornadic waterspouts. Unlike the tornadic variety, fair-weather waterspouts can arise independently of **severe thunderstorm**s. They may appear singly or in clusters. Fair-weather waterspouts are seldom energetic enough to cause any damage, except to small craft directly in their path.

Fair-weather waterspouts are usually much smaller than the tornadic variety, with diameters ranging from 10 to 325 feet (3 to 100 me-

Tornadoes

WEATHER REPORT: VIEWING A FUNNEL UP-CLOSE

Will Keller is one of the few people who has looked directly into a tornado's funnel and lived to tell about it. Keller was a farmer who lived in Kansas. On June 22, 1928, he was working in a field when he saw a severe storm approaching. He delivered the following account at the Weather Service office in Dodge City, Kansas:

"On the afternoon of June 22, 1928, I was out in my field with my family looking over the ruins of our wheat crop which had just been completely destroyed by a hailstorm. I noticed an umbrella shaped cloud in the west and southwest and from its appearance suspected that there was a tornado in it. The air had that peculiar oppressiveness which nearly always precedes a tornado.

I saw at once that my suspicions were correct, for hanging from the greenish-black base of the cloud were three tornadoes. One was perilously near and apparently heading for my place. I lost no time hurrying my family to our cyclone cellar.

The family had entered the cellar and I was in the doorway just about to enter and close the door when I decided to take a last look at the approaching twister...."

ters). The speed of their rotating winds approaches 50 mph (80 kph). They tend to last a mere ten to fifteen minutes and move along the water at a slow pace.

Sometimes waterspouts are spawned by small thunderstorms. Most often, however, they occur beneath **Cumulus congestus** clouds (puffy, cauliflower-shaped clouds), the tops of which usually remain below 12,000 feet (3,600 meters).

Fair-weather waterspouts develop in a process that is similar to the formation of a **thunderstorm.** Namely, both weather systems require an **unstable** atmosphere and warm, humid air at the surface.

In waterspouts, converging winds, possibly due to **sea breezes** or

> There was little or no rain falling from the cloud. Two of the tornadoes were some distance away and looked like great ropes dangling from the parent cloud, but the one nearest was shaped more like a funnel, with ragged clouds surrounding it....
>
> Steadily the tornado came on, the end gradually rising above the ground.... At last the great shaggy end of the funnel hung directly overhead. Everything was as still as death. There was a strong, gassy odor and it seemed as though I could not breathe. There was a screaming, hissing sound coming directly from the end of the funnel. I looked up and, to my astonishment I saw right into the heart of the tornado. There was a circular spinning in the center of the tornado, about 50 to 100 feet in diameter, which extended straight up for a distance of at least one-half mile, as best I could judge under the circumstances. The walls of this opening were rotating clouds and the whole was brilliantly lighted with constant flashes of lightning which zig-zagged from side to side....
>
> Around the lower rim of the great vortex, small tornadoes were constantly forming and breaking away. These looked like tails as they writhed their way around the end of the funnel.... The opening was entirely hollow, except for something I could not exactly make out—perhaps a detached wind cloud—that kept moving up and down. The tornado was not traveling at a great speed so I plenty of time to get a good view whole thing, inside and out."

gust fronts, trigger the surface air to rise. As the air rises, it cools, and the moisture within it condenses into cloud droplets. The **condensation** releases **latent heat** which adds to the instability of the surrounding air and enables the moist air from the surface to continue rising. Intense **updraft**s then form within the cloud.

A waterspout funnel, similar to a tornado funnel, is comprised of updrafts rotating about a **vortex.** And like some tornado funnels, the waterspout is made visible by condensation of moisture within the rising air.

As with the swirl of debris beneath a developing funnel cloud, the first clue to a forming waterspout is often swirling winds just above the surface of the water. Once fully formed, a waterspout takes on the shape

Tornadoes

A KEY REFERENCE TO: TORNADO ALERTS AND SAFETY PROCEDURES

A tornado **watch** means that weather conditions are favorable for the development of **tornado**es in your area. Stay to tuned to your radio or television for updates and prepare to move quickly to a safe place. Watch the sky, as well. You may spot the warning signs of a tornado (see section on approaching tornadoes, page 254) before a tornado warning has even been issued.

A tornado **warning** means that a tornado has been detected in your area, either visually or by weather **radar.** Television and radio programs are interrupted, and in many communities sirens sound, to announce a tornado warning. When a tornado warning is issued, it's time to move to a safe place immediately. You may only have seconds before a tornado strikes.

In preparation for a tornado strike, it's wise to have areas designated as "tornado shelters" at home, at school, and at work. The best place for a tornado shelter is in a basement. If there is no basement, select an interior room (bathrooms and closets are best) or hallway on the first floor, away from windows. Store a first-aid kit and a flashlight with extra batteries in your tornado shelter.

When a tornado warning is issued:
- Go to your designated tornado shelter. Crouch beneath the stairs, a heavy workbench, a mattress, or a sturdy piece of furniture.

of an arc, stretching from the cloud to the surface of the water. It dissipates when cool air gets drawn into the funnel.

Waterspouts are common in coastal areas of all tropical oceans. They can also form over large inland bodies of water. For instance, they occur from time to time in the summer on the Great Lakes and on Utah's Great Salt Lake. They are also a common sight over the Mediterranean Sea. The Florida Keys stand alone, however, as the waterspout capital of the world. Throughout the summer, almost 100 fair-weather waterspouts occur there per month. Waters around the Keys see a total of 400 to 500

- Do not open windows! It was once believed that opening windows would preserve structures by allowing indoor and outdoor pressure to equalize. It is now known that opening windows only increases pressure on the opposite wall, making it more likely that the building will collapse.
- If you are outside, go into a strong building. Take shelter away from windows and doors. Flying glass and other debris are major tornado hazards.
- If you are in a car, leave the car and go into a nearby building. If that's not possible, either because you are far from buildings or the building is locked, leave your car and crouch in a ditch or depression or beside a strong building. Do not stay in the car! Tornadoes can pick up cars and hurl them through the air. When the car is dropped to the ground, it may crash with the force of a 100-mph (160-kph) head-on collision.
- If you are in a mobile home, leave! Even properly secured mobile homes can be lifted up by a tornado. Go to a designated tornado shelter or crouch in a ditch.
- Lend assistance to very young children, elderly people, people who are mentally or physically disabled, and people who don't understand the tornado warning due to a language barrier.
- Protect yourself by lying face down with your knees drawn up under you. Put your head down by your knees and cover the back of your head with your hands.

waterspouts per year. The exact number of waterspouts is hard to determine, since many of those that cause no damage go unreported.

The Florida Keys are conducive to waterspout formation due to three factors: heat, humidity, and the **trade winds.** First, the islands and the shallow water surrounding them are heated to temperatures ranging from the mid-80s to the low 90s, in °F (about 30 to 35°C). The land and water transfer heat to the air by **conduction.** The surface air, which has a high humidity, rises and forms clouds.

The clouds are then blown into relatively straight lines by the trade winds, which blow from the east or northeast. For a yet-to-be discovered reason, lines of clouds enhance the formation of waterspouts.

HURRICANES

Hurricanes certainly deserve the title "greatest storms on Earth." After all, no other storm system has the power of a hurricane. One might argue that a **tornado** is mightier, since tornadoes have faster winds. While it is true that the winds of a tornado can reach speeds twice those of a hurricane, a tornado is a tiny storm in comparison to a hurricane.

A relatively small hurricane is 100 miles (160 kilometers) in diameter, while the diameter of one of the largest tornadoes is no more than about a mile. And while tornadoes travel for less than 10 miles (16 kilometers), on average, and up to 200 miles (320 kilometers) at their greatest extent, hurricanes regularly travel for thousands of miles. Tornadoes usually last for only minutes, and occasionally as long as a few hours. Hurricanes, in contrast, often last over a week.

To demonstrate the power contained in a hurricane, consider the following: 1) 1 percent of the energy in an average hurricane could supply the entire United States with power for one year; and 2) the energy unleashed by a hurricane in one day is equal to the energy of four hundred 20-megaton hydrogen bombs.

Hurricanes are among the deadliest and most destructive of all natural disasters. A single strong hurricane can kill hundreds of people. Fortunately, the number of deaths caused by hurricanes has declined dramatically in recent years, due to the development of early detection systems. In the last five years, hurricanes have accounted for about 130 deaths in the United States. At the same time, however, due to population growth in hurricane-prone areas, the cost of property damage by hurricanes has been steadily rising.

WHAT ARE HURRICANES?

A hurricane is the most intense form of **tropical cyclone.** A tropical cyclone is any rotating weather system that forms over tropical waters. As discussed in "What Is Weather?" winds blow inwards toward, and rotate around, an area of low-pressure. Winds rotate counterclockwise in the **Northern Hemisphere** and clockwise in the **Southern Hemisphere.** To qualify as a hurricane, a storm must have a well-defined pattern of rotating winds and maximum sustained winds greater than 74 mph (119 kph).

A hurricane is made up of a series of tightly coiled bands of **thunderstorm cloud**s. These bands spiral around an almost totally calm circle, called the **eye,** at the center of the hurricane. There may be hundreds of strong **thunderstorm**s within a hurricane. The diameter of an average hurricane is about 350 miles (560 kilometers) while the diameter of the largest hurricanes approaches 900 miles (1,500 kilometers).

Less intense forms of tropical cyclones are referred to as **tropical storm**s, **tropical depression**s, or **tropical disturbance**s. A tropical storm is similar to a hurricane in that it has organized bands of rotating strong thunderstorms, yet its maximum sustained winds are only 39 to 73 mph (63 to 117 kph). A tropical depression, the weakest form of tropical cyclone, consists of rotating bands of clouds and thunderstorms with maximum sustained winds of 38 mph (61 kph) or less. A tropical distur-

A hurricane photographed from space by an astronaut.

Hurricanes

bance is a cluster of thunderstorms that is beginning to demonstrate a cyclonic circulation pattern.

A hurricane starts out as a tropical disturbance and passes through the stages of tropical depression and tropical storm on its way to maturity. A hurricane passes back through those stages, in the reverse order, as it dissipates.

Over the ocean, a hurricane generates waves 50 feet (15 meters) or greater in height. And when a hurricane reaches land, it pounds the shore with a wall of water up to 20 feet (6 meters) high. This wall can produce severe flooding across 100 miles (160 kilometers) of coastline. Hurricanes also bring fierce winds and intense downpours of rain. It is not unusual for coastal and inland communities to receive 6 to 12 inches (15 to 30 centimeters) of rain when a hurricane comes onshore. For its onshore finale, a dissipating hurricane may spin off numerous tornadoes.

The word "hurricane" is used to refer to tropical cyclones that form in the northern Atlantic Ocean, Caribbean Sea, Gulf of Mexico, or in the eastern Pacific Ocean off the coasts of Mexico and Central America. Hurricanes that occur in the western North Pacific and China Sea region are called "**typhoon**s." The word "typhoon" comes from *tai-fung,* which is Cantonese for "great wind." Hurricanes that form over the Indian Ocean are called "**cyclone**s." Hurricanes are called "baquiros" in the Philippines, "willy-willies" in northwest Australia and the Timor Sea region, and "huracans" in the West Indies.

Figure 31: Air flow within a hurricane.

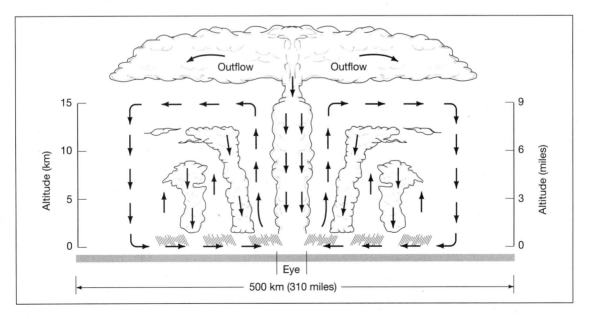

Structure of Hurricanes

As stated above, a hurricane consists of spiraling bands of clouds called **rain band**s, around a calm, low-pressure center, or eye. The rain bands, which are tightly coiled around the eye, produce heavy rains and forceful winds. Surrounding the outer edge of the rain bands is a region of wispy, high-level **cirrus** or **cirrostratus** clouds.

The hurricane's eye has an average diameter of 12 to 40 miles (20 to 65 kilometers). Within the eye, winds are light and skies are partly cloudy-to-clear. The reason that clouds break up in the eye of a storm is that air sinks in that region. Air warms as it falls and moisture within it evaporates.

The lowest pressure of the storm exists in the eye. Typically, pressure there dips to 950 millibars (28 inches), although measurements as low as 900 millibars (26.5 inches) have been recorded. By way of comparison, the average pressure at sea level is 1,013.25 millibars (29.92 inches).

It may take an hour or more for the eye of the storm to pass over an area. The calm weather associated with the eye sometimes fools the residents of that area into thinking the storm is over when, in fact, the heavy winds and rain will soon resume.

The region immediately surrounding the eye, called the **eye wall,** is the strongest part of the storm. The eye wall is a loop of thunderstorm clouds that produce torrential rains and forceful winds. The closer one gets to the center of the storm without actually entering the eye, the faster the winds blow. Within a radius of 6 to 60 miles (10 to 100 kilometers) of the eye, winds may reach speeds of 100 to 180 mph (160 to 300 kph).

The winds are driven by the **pressure gradient** between the edge of the storm and the eye. The closer in to the eye, the steeper the pressure gradient becomes and the faster the winds blow. At points farther away from the eye of the storm, the pressure gradient becomes more gradual and the winds become weaker.

The most violent part of the hurricane is the side of the eye wall in which the wind blows the same direction that the storm is progressing. In that region, the hurricane's winds combine with the winds that are steering the hurricane, to create the storm's fastest winds.

The cyclonic wind circulation of the hurricane weakens with height, starting at about 9,800 feet (3,000 meters). The airflow actually reverses direction at heights greater than approximately 50,000 feet (15,000 meters). That means that while the eye of the storm and the surface beneath it are intensely low-pressure areas, a high-pressure area exists above.

Hurricanes

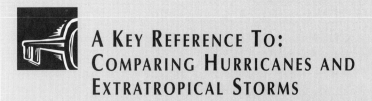

A Key Reference To: Comparing Hurricanes and Extratropical Storms

A tropical cyclone, the strongest variety of which is a hurricane, forms under a wholly different set of circumstances from those that produce an **extratropical cyclone.** An extratropical cyclone is a large-scale storm that occurs in the middle latitudes (see "What Is Weather?" on page 41). Whereas an extratropical cyclone is produced by collisions between contrasting **warm** and **cold fronts,** a hurricane often develops in the absence of fronts, and always within a single **air mass.**

The issue of fronts is particularly relevant regarding each storm system's source of energy. An extratropical cyclone is fueled by temperature contrasts between the air masses found on either side of a front; a hurricane is fueled by the warmth and moisture of the ocean and the **latent heat** released as the surface air rises.

And whereas both extratropical cyclones and hurricanes are centered around an area of low-pressure, a hurricane has a much steeper pressure gradient of the two systems. And typically the air pressure at the center of a hurricane is much lower than it is at the center of an extratropical cyclone.

Other differences between the two types of cyclones include:

The fact that a hurricane has a ceiling below the top of the **troposphere** makes it possible to fly above a hurricane in an aircraft and take aerial photographs of the entire system. It is not possible to do this in a **middle-latitude** thunderstorm since the mature **cumulonimbus** clouds in that system extend to the top of the troposphere and sometimes beyond.

Hurricanes Are Tropical Phenomena

The tropics are defined as the region of Earth bounded by 23.5 degrees **latitude,** North and South. The tropics receive the most direct sunlight of anywhere on the planet, making that region the world's warmest.

- The diameter of the typical extratropical cyclone is three times the size of the diameter of the typical hurricane.
- Hurricanes weaken with height while extratropical cyclones intensify with height.
- Air sinks in the center of a hurricane while it rises in the center of an extratropical cyclone.
- Air at the center of a hurricane is warmer than the surrounding air while the air at the center of an extratropical cyclone is colder than the surrounding air.
- The winds of a hurricane are strongest at the surface while the winds of an extratropical cyclone are strongest at upper levels, in the **jet stream.**

It is important to note that, despite all these differences, hurricanes that move inland occasionally develop into extratropical cyclones. This change occurs when a hurricane moves across a **front** and draws in air of different temperatures. Then the hurricane, which was weakening as it traveled over land, re-intensifies as it becomes linked with a low-pressure area aloft.

An example of a hurricane that became linked with an extratropical storm is Hurricane Agnes of 1972. After combining with a low-pressure system in the northeast, Agnes produced **heavy rain**s and extensive flooding. Harrisburg, Pennsylvania, for instance, received 12.5 inches (32 centimeters) of rain in a twenty-four-hour period. The rising waters in central Pennsylvania forced the evacuation of more than 250,000 people. Agnes caused a total of 122 deaths and $6.4 billion in property damage. The **flood**s were responsible for around half of the deaths and about two-thirds of the property damage.

The heating of Earth's surface leads to the daily formation of **cumulus** clouds and afternoon thunderstorms. While these thunderstorms are generally not severe individually, they sometimes become organized in lines called **tropical squall cluster**s or squall lines, which are severe. These are similar to **squall line**s of thunderstorms that form over land in the middle latitudes (see "Thunderstorms," page 223).

Hurricanes form only within specific areas in the tropics, namely between 5 degrees and 20 degrees North and South, although they occa-

Hurricanes

sionally form as far north and south as 30 degrees. At higher latitudes, the water is too cold for hurricanes to form.

The reason why hurricanes won't form between 5 degrees North and 5 degrees South has to do with the lack of **Coriolis effect** at and near the equator. The Coriolis effect is the bending of global winds due to the rotation of Earth (see "What Is Weather?" on page 20). The Coriolis effect is necessary for the formation of rotating a tropical storm. And at the equator, the **trade winds** from the north and south meet and the Coriolis effect is canceled out.

FORMATION OF HURRICANES

The first step in the formation of a **hurricane** is the development of a cluster of **thunderstorm**s, called a **tropical disturbance.** Tropical disturbances develop regularly over tropical waters. Only a small percentage of these disturbances, however, evolve into hurricanes.

According to a study conducted by examining satellite photos, only 50 of the 608 tropical disturbances detected in a six-year period over the Atlantic Ocean grew into **tropical storm**s. It has also been shown that half to two-thirds of all tropical storms develop into hurricanes. Thus, only about 4 to 6 percent of all newly created storm systems in the tropics develop into hurricanes.

The reason that hurricanes are not more common is that hurricane formation requires a very specific set of atmospheric conditions.

HURRICANE INGREDIENTS

Some of the factors necessary for hurricane formation are similar to those necessary for thunderstorm formation. These include a warm, moist **air mass** and an **unstable** atmosphere. In the case of a hurricane, the air is warmed and made moist by ocean water that is at least 80°F (27°C). And since ocean water is stirred up by storms, this water must be warm to a depth of about 200 feet (61 meters).

Hurricane formation also requires that the air, from the surface up to about 18,000 feet (5,500 meters), be extremely humid. The higher the temperature of the ocean water, the more water evaporates. That, in turn, raises the humidity of the surface air. The surface air cools as it rises and the moisture within it condenses. As we described in the chapter on "Thunderstorms," when moisture condenses it releases **latent heat,** which provides energy to the storm system.

Another necessary ingredient of a hurricane is surface winds that are converging, or blowing toward a common point. Where winds converge at the surface, air rises. A number of events can trigger surface air **convergence.** One is that a **front** moves into the tropics from the **middle latitudes.** By the time the front reaches the tropics, the temperatures on either side of the front have equalized. However, the front still has an associated low-pressure area aloft. The winds at the surface converge to a point beneath that low-pressure area and rise.

Another source of convergence of surface winds is the meeting of the **trade winds** from the north and the south, along the **intertropical convergence zone,** or ITCZ (see "What Is Weather?" on page 25). Convergence at the ITCZ promotes hurricane development only during the of winter and summer when the ITCZ is located furthest from the equator. At other times of year, when the sun is directly over the equator, the ITCZ also sits very close to the equator. At those times the ITCZ does not experience the **Coriolis effect.** Once the ITCZ shifts to a point 4 or 5 degrees from the equator, however, the Coriolis effect is apparent.

A third possible source of converging winds is an easterly trade wind that flows in a wavelike fashion from Africa. As the trade wind proceeds westward, a low-pressure **trough** may flow over the tropical waters, triggering surface winds to converge beneath it.

Hurricane formation also requires ideal conditions in the **winds aloft.** Namely, the winds at all altitudes need to be light and blowing in approximately the same direction and speed, so as not to scatter the moisture and dissipate the developing storm. Hurricane formation is most likely to occur when the upper air is cold, a factor that contributes to the low pressure.

Most often, there is some condition in the winds aloft that prohibits hurricane development. For instance, in some **latitude**s, particularly between 20 and 30 degrees, the upper-level air sinks. It warms as it falls, creating an **inversion** that prevents the upward development of thunderstorms.

Another condition that works against hurricane formation is a relatively dry middle layer of air. The amount of latent heat necessary to generate a hurricane requires **condensation** to occur at all heights throughout the system. And moisture will not condense until the **relative humidity** of the air has reached 100 percent.

Hurricanes

How Hurricanes Unfold

Let's assume that the following atmospheric conditions exist: warm, humid surface air; cold air and low pressure aloft; converging surface winds; and light winds and high humidity at all altitudes. The stage has been set for the development of a hurricane.

The most widely accepted model of how a hurricane forms is called the **organized convection theory.** According to this theory, the first step in the development of a hurricane is the formation of large **thunderstorm cloud**s. That proceeds as follows: As warm air rises to the region of low pressure above, condensation occurs. Latent heat released during condensation warms the air at greater and greater heights. Eventually, the cold layer of upper air also becomes warm. As a result, its **air pressure** increases.

When the low-pressure region aloft becomes transformed into a high-pressure region, the thunderstorm ceases to develop upward.

Figure 32: Formation of a hurricane according to the organized convection theory.

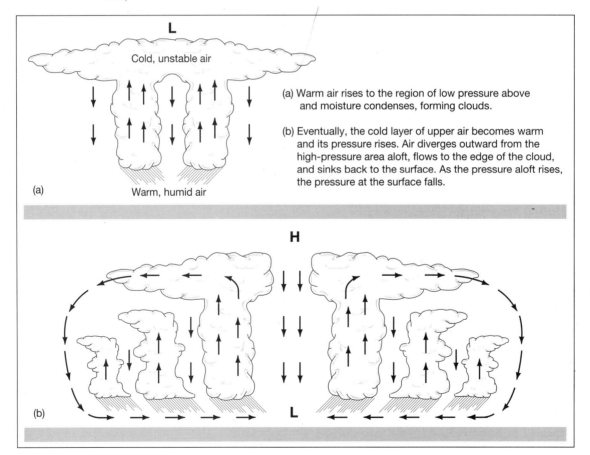

Rather, at that height air **diverges** outward, away from the high-pressure area at the center of the cloud. The air flows to the edge of the cloud and then sinks back to the surface.

As the same time that the pressure aloft rises, the pressure at the surface falls. And it is the formation of the surface low, at the center of the storm, that is the most essential factor in hurricane development. The winds begin to circulate in a counterclockwise pattern (in the **Northern Hemisphere**) around the center of, but not directly into, this surface low. At the center of the storm, the air sinks directly from the high pressure area above to the low pressure, below.

Winds increase in speed the closer they get to the center. In the process, they generate large ocean waves. These waves create friction on the wind, interrupting the air flow and causing the air to converge. Where the air converges, it rises, carrying moisture and warmth upward to form new thunderstorms.

The rising air increases the air pressure aloft. As a result, the surface air pressure becomes even lower. A chain reaction is set in motion: new thunderstorms are formed, the upper-level high-pressure area becomes higher, and the surface low-pressure area becomes lower. Each process drives the other as the storm grows into a mature hurricane.

A hurricane continues to grow as long as a fresh supply of warm, humid air is available. Once the hurricane crosses over colder waters or land, its source of energy is cut off, and it begins to dissipate.

As long as more air flows out from the top of the storm center than flows into the storm from the surface, the hurricane continues to intensify. The point at which that trend reverses the hurricane begins to die.

The level of air flow is a indication of air pressure. A rapid outflow at the top of the hurricane indicates that pressure is high at upper levels and, consequently, low at the surface. However, when the outflow slows, the pressures at the top and bottom of the storm center begin to equalize. In the absence of a strong surface low, the winds weaken. Then there is nothing to support the coiled organization of thunderstorms. The system unwinds and individual thunderstorms dissipate.

LIFE CYCLE OF HURRICANES

A typical hurricane lasts about thirteen days. It starts out on day one as a cluster of thunderstorms, a condition that meteorologists call a **tropical disturbance.** By day three, the thunderstorms have become orga-

Hurricanes

nized into bands that swirl about a low-pressure center. The system, at this point, is called a **tropical depression.**

The surface air pressure drops, the winds intensify, and the storm continues to grow. By day five, winds are blowing faster than 39 mph (63 kph). At this stage, the system is upgraded to a **tropical storm** and given a name. By day seven winds are blowing faster than 74 mph (118 kph). The storm is now classified as a hurricane.

For the next few days, as the hurricane moves across the warm water, it maintains its strength and integrity. About day twelve it crosses onto land and weakens. By day thirteen it has dissipated.

Distribution of Hurricanes

Hurricane-breeding areas are found in several clusters within the world's tropical oceans. Of equal importance to where **hurricane**s form is the path along which they travel. In this section we will look at where and when hurricanes form and examine what shapes their paths.

Where Hurricanes Form

Hurricanes form over only certain parts of the world's tropical oceans. The reason that they don't form over all tropical waters is that in certain areas the water temperature is not high enough. For instance, the

Figure 33: The World's hurricane breeding regions.

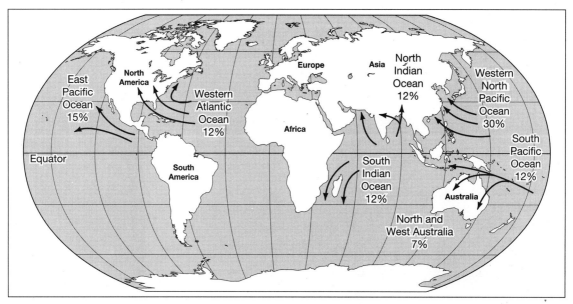

south Atlantic and southeastern South Pacific oceans on either side of South America are notably hurricane-free. Figure 33 shows the primary hurricane-breeding regions.

WHEN HURRICANES FORM

The annual hurricane season, for any given location, is during the months when ocean temperatures are highest. This period lags behind the year's warmest months on land, since it takes the oceans longer than the land to both warm up and cool down (see "What Is Weather?" on page 70).

For the **Northern Hemisphere,** hurricane season is roughly June through November. In the **Southern Hemisphere,** hurricanes occur most frequently between December and May. The exception to this rule is the western portion of the north Pacific Ocean, where hurricanes form year-round.

Within the long hurricane season, there are peak months for hurricane formation that vary with location. For instance, the maximum number of hurricanes in the north Atlantic region—hurricanes that threaten the U. S. eastern seaboard—occur in August and September.

HOW HURRICANES MOVE IN OUR PART OF THE WORLD

Hurricanes that form over the north Pacific and north Atlantic oceans are guided to the west-northwest by easterly **trade winds.** These winds blow a hurricane along at about 10 mph (16 kph). Once a hurricane encounters the Azores-Bermuda High, the **semipermanent high pressure** system in the east Atlantic (see "What Is Weather?" on page 28), it is directed to the northwest, through the Caribbean and toward the East Coast of the United States.

If the hurricane travels so far northward that it reaches the **middle latitude**s, it will be steered to the northeast by the **westerlies.** The westerly winds blow the hurricane along at about 55 mph (88 kph), which is significantly faster than the trade winds. This faster movement at higher latitudes makes it more challenging to get advance warning to communities in the hurricane's path.

While the trade winds, Azores-Bermuda High, and westerlies establish a general hurricane route, the specific path taken by a hurricane is much more difficult to predict. The specific path depends on the structure of the hurricane and how it interacts with its environment. Exactly how these factors influence hurricane movement is not well understood. The path also depends on the size and location of the Azores-Bermuda High, which vary over the course of the year.

Hurricanes

While some hurricanes follow a smooth course, others travel erratically, shifting direction suddenly and inexplicably. For instance, some hurricanes that seem certain to spare the U.S. Atlantic Coast will turn to the west and crash onto shore. On the other hand, coastal communities may brace for a hurricane's onslaught only to be spared at the last minute as the storm turns back to sea.

Hurricanes that form over the eastern Pacific, off the west coast of Mexico, generally move to the west or northwest and travel over the ocean until they dissipate. That is the reason we hear so little about those hurricanes. Occasionally, however, a Pacific hurricane will turn to the north or northeast and strike the west coast of Mexico. A handful of such hurricanes have devastated Mexican coastal communities over the past several decades. The remnants of those hurricanes have brought heavy rains and flooding to the U.S. Southwest and West Coast.

One would expect Hawaii to suffer from a large number of hurri-

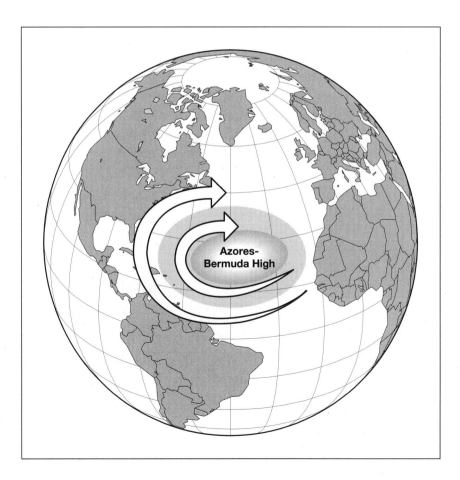

Figure 34: Direction of hurricane movement in the North Atlantic.

canes, given that Hawaii lies in the path of hurricanes formed off the west coast of Mexico. But that is not the case. As it turns out, Hawaii is located far enough to the west that by the time most hurricanes reach it, they have been significantly weakened.

There have been exceptions to this rule, however. About once a decade, Hawaii is struck by a major hurricane. A notable hurricane in recent history is Hurricane Iniki, which struck the island of Kauai on September 11, 1992. Iniki caused $1.8 billion worth of damage and seven deaths.

HURRICANES THAT STRIKE THE UNITED STATES

The only hurricanes that directly affect the United States are those formed over the tropical north Atlantic, the Caribbean, or the Gulf of Mexico. On average, six hurricanes form over those waters during each hurricane season, two or three of which strike the U.S. Atlantic or Gulf

Hurricanes

Commercial buildings in downtown Kapaa rest with their roofs split open in the aftermath of Hurricane Iniki. Most of the business section of the town was destroyed.

Hurricanes

EXTREME WEATHER: TRACKING GILBERT

On September 10, 1988, when Tropical Storm Gilbert was just south of Puerto Rico, it was upgraded to a hurricane. Gilbert intensified as it headed west-northwest, passing to the south of the Dominican Republic and Haiti on September 12. It then blasted right through Jamaica on September 13, killing 45 people.

Hurricane Gilbert continued to strengthen the next day, as it headed to the northwest, toward Mexico's Yucatan Peninsula. On September 15 Gilbert ripped through the Peninsula's resort towns with steady winds of 175 mph (280 kph). Gilbert traveled across the Gulf of Mexico and on the 17th struck the Mexican coast, just south of Texas.

As Gilbert continued to the northwest, across northern Mexico, it began to weaken, but not before it had killed 202 people. On September 18, Gilbert began to veer to the northeast. As Gilbert moved across Texas on September 18 and 19, it weakened further. In its last stages, Gilbert spawned 29 **tornado**es in Texas over a 33-hour period.

Damage due to Hurricane Gilbert totaled over 315 fatalities, tens of thousands rendered homeless, and around $5 billion in economic damage.

coasts. The primary factor that determines whether or not a hurricane hits the United States is the location at which the hurricane moves to the northeast. If a hurricane is still over the ocean when it changes course from northwest to northeast, it will miss the coast. However, some hurricanes move over land before they change direction.

The part of the United States that experiences the greatest number of hurricanes is the Florida Keys. The Keys have been hit by 43 hurricanes in the last 100 years. The places where hurricanes pose the greatest threat to humans, however, are the barrier islands that exist along portions of the Atlantic Coast and Gulf Coast. Evacuation from those islands poses a much greater challenge than evacuation from mainland coastal areas. Adding to the difficulty in recent years has been the explosion of development on some of these islands. And the pace of the construction of evacuation routes, overall, has not kept pace with the population growth.

The problem of evacuation from the barrier islands, however, goes back many years. For example, in 1893 thousands of freed slaves who had homesteaded on the South Carolina sea islands were killed in a hurricane. And in 1900, around seven thousand residents of Texas' Galveston Island perished in the what was the worst storm of this century (see box, page 281).

For reasons discussed above, the U.S. West Coast is rarely struck by hurricanes. Once in every several years, a **tropical storm** or hurricane does strike. A recent example occurred on September 9–10, 1976, when Hurricane Kathleen traveled across Baja California and brought intense rains and floods to Arizona.

DESTRUCTIVE FORCES OF HURRICANES

When a **hurricane** moves onto land, it is capable of causing tremendous damage. While a hurricane's winds are often thought to be its most destructive element, this is not the case. While the winds do cause a great deal of damage, **flood**s caused by ocean swells and torrential rains cause the most hurricane damage. Floodwaters, both in coastal and inland areas, account for about 90 percent of hurricane fatalities. An-

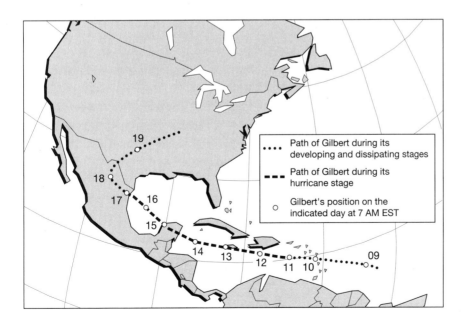

Figure 35: The path of Hurricane Gilbert.

Hurricanes

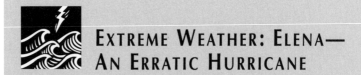

EXTREME WEATHER: ELENA— AN ERRATIC HURRICANE

Some hurricanes change direction several times. Such was the case with Hurricane Elena in August 1985. Elena traveled into the Gulf of Mexico and then veered toward the Florida Panhandle. This change prompted the evacuation of the Panhandle's coastal communities. Just before striking land, Elena turned to the east and headed for Tampa Bay. The residents there were then evacuated.

Since they seemed to be out of danger, the Panhandle residents returned home. Just as they were settling in, however, Elena turned back in their direction. The people in the Panhandle communities were forced to leave their homes again, along with people living on the coasts of Alabama, Mississippi, and Louisiana. Elena finally came onshore near the border of Mississippi and Louisiana and caused around $1.4 billion worth of damage.

other destructive force of hurricanes is the **tornado**es that are spawned by the storms in their final stages.

STORM SURGES

In coastal areas, most of the property damage, as well as about 90 percent of deaths, are due to the hurricane's **storm surge.** A storm surge is a wall of water that sweeps on shore when the **eye** of a hurricane passes overhead. Storm surges range from 3 to 6.5 feet (1 to 2 meters) in a weak hurricane, to over 16 feet (5 meters) in a strong hurricane. A storm surge affects a stretch of coastline between 40 to 100 miles (65 to 160 kilometers) long. It levels any structure in its path.

The abnormal swelling of the ocean that produces a storm surge is caused by the combined effects of a storm's **pressure gradient** and high winds. In the open ocean, strong winds and high pressure around the edges of the hurricane push down on water, lowering its level. The displaced water flows toward the center of the storm, where pressure is lowest. The water at the center rises and then spirals downward to about 200 feet (60 meters) beneath the surface, where it flows outward.

When the storm moves into shallow waters, there is no place for the mound of water beneath the storm center to descend. Thus, the water there is forced to pile upward. The mound of water reaches its greatest height where it crashes onto shore. The storm surge is highest in strong hurricanes, especially where the ocean floor slopes gradually to the shore.

The storm surge is equal to the height of the water over the normal tide level. Consider the case in which a hurricane makes **landfall** at high tide. The term landfall refers to the point at which a hurricane crosses from the ocean onto land. Say the water at high tide is normally 2 feet above sea level. If a hurricane produces a wall of water 12 feet above sea level, then the storm surge is 10 feet. The storm surge plus the tide, which in this case is a 12-foot-tall wall of water, is called the **storm tide.**

The largest storm tide in recent U.S. history was 25 feet (8 meters). It was caused by Hurricane Camille in August 1969, which made landfall in Pass Christian, Mississippi. The storm tide destroyed more than 5,500 homes. In addition, it damaged around 12,500 homes and 700 businesses. The area of greatest destruction was along a 60-mile-long (97-kilometer-long) area of coastline in Mississippi, Alabama, and Louisiana. And as far as 125 miles away, the water level was 3 feet higher than usual.

The largest recorded storm surges have occurred in other parts of the world. In 1737, a storm surge estimated to be 40 feet (12 meters) tall

A hurricane howls on the South Sea island of Manakoora in the 1979 re-make of John Ford's 1937 classic film, Hurricane.

Hurricanes

struck the Bay of Bengal, killing more than 300,000 people. And a storm surge of 42 feet (13 meters) inundated Bathurst Bay, Australia, in 1899.

The destructive capabilities of a storm surge are the result of two factors: the density of the sea water, which is about 64 pounds (141 kilograms) per cubic foot; and the debris that the water sweeps along. Examples of debris include boats that were ripped from their moorings by the waves, and pieces of destroyed buildings, trees, and sand. As the water plus this debris moves farther onshore, it batters and flattens anything in its path. Another way that a storm surge destroys a coast is that it erodes sand and soil. When sand and soil is washed away beneath buildings, roads, and sea walls, these structures buckle or collapse.

A storm surge may also have the secondary effect of causing flooding of inland bays and rivers. This flooding results when storm surge water is squeezed into narrow channels. Sometimes inland waters rise even higher above normal levels than coastal waters in the wake of a storm surge.

WINDS

The winds of a hurricane also cause a significant amount of damage to coastal areas. As we stated earlier in this chapter, hurricane winds range between 75 and 180 mph (120 and 290 kph). Winds of this strength can damage buildings and homes and knock down trees and telephone poles, as well as cause beach erosion. And they can totally demolish

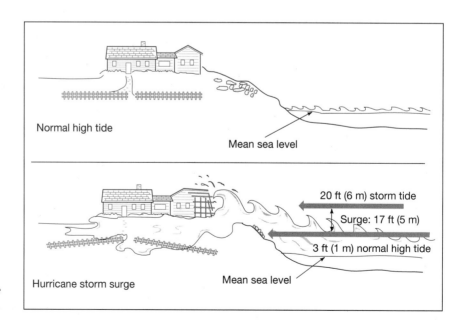

Figure 36: A hurricane storm surge.

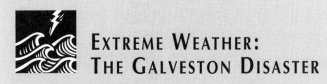

Extreme Weather: The Galveston Disaster

The greatest hurricane disaster and, indeed, the greatest weather disaster, in the history of the United States occurred on September 8, 1900, in Galveston, Texas. More than 6,000 people in that low-lying barrier island town on the Gulf of Mexico lost their lives in a 20-foot-high (6-meter-high) storm surge.

Galveston's residents had received warning from the U.S. Weather Bureau on September 6 that a **tropical storm** was detected near Cuba. However, they did not pay much notice. After all, other tropical storms had come their way and had inflicted only minor damage. Galveston residents had no way of knowing the power of the storm that was about to strike them.

It was not until the morning of September 8 that the higher-than-usual tides and strong winds hinted at the severity of the approaching storm. However, rather than evacuating the island, most residents merely took shelter in brick houses at higher elevations. The problem with that strategy was the highest point on the island, at that time, was only 9 feet (about 3 meters) above sea level. That meant that no place on the island was safe from the storm's rushing waters.

The hurricane's assault on Galveston lasted for several hours, starting at about 4 P.M. Winds gusting to 100 mph (160 kph) were recorded before the island's **anemometer** was destroyed. A 20-foot (6-meter) storm surge crashed onto shore and entirely submerged the island. To make matters worse, the hurricane also produced 10 inches (25 centimeters) of rain. Once the storm had passed and flood waters subsided, very few structures were left standing in the town.

Since that time, Galveston has been rebuilt and extensive measures have been taken to prevent against future disasters. The residents have constructed a 17-foot-tall (5- meter-tall), 3-mile-long (5-kilometer-long) seawall facing the Gulf of Mexico. They also brought in sand and raised the elevation of the island, at some points as high as 17 feet (5 meters) above sea level.

Hurricanes

Hurricane Andrew: The Costliest Natural Disaster in U.S. History

In August 1992, Hurricane Andrew struck Florida and Louisiana, causing 58 deaths and $30 billion in damage. Over 200,000 homes and business were damaged or destroyed and 160,000 people became homeless. Andrew, which made landfall as a category 4 hurricane, was the costliest natural disaster in the history of the United States.

On August 21, Andrew, which was then classified as a **tropical storm** in the Atlantic, appeared to be weakening. However, it then moved over warmer waters and rapidly gained strength. On August 21, its winds were only 52 mph (84 kph). Two days later, its winds had increased to 140 mph (225 kph), and Andrew had developed into an intense hurricane.

Andrew came on shore at Homestead, on the southern tip of Florida, on August 24. With winds that peaked at about 200 mph (320 kph) and a 16.9-foot-tall storm surge (a record for Florida), Andrew devastated Homestead. It leveled trees, utility poles, and 50,000 homes.

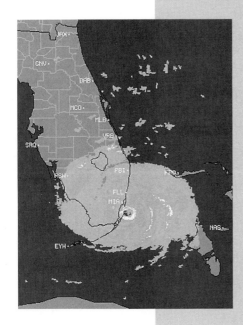

Andrew then traveled westward, over land, and into the Gulf of Mexico. While Andrew had weakened during its journey across the land, it regained strength over the Gulf's warm waters. On August 25, Andrew blew into Louisiana with 138-mph (222-kph) winds. There, Andrew continued its relentless destruction of property.

Andrew's death toll was relatively low due to the efficiency of prediction and warning systems. Over 1 million people in Florida and 1.7 million people in Louisiana and Mississippi were evacuated from areas in the storm's path. Had Andrew occurred in the first half of this century, before the development of sophisticated hurricane-detecting technology, the death toll would have certainly been much higher.

Ten Deadliest Hurricanes to Strike the United States Since 1900

Hurricane	Year	Category	Deaths
Texas (Galveston)	1900	4	6,000
Florida (Lake Okeechobee)	1928	4	1,836
Florida (Keys and S. Texas)	1919	4	600–900
New England	1938	3	600
Florida (Keys)	1935	5	408
Audrey (Louisiana and Texas)	1957	4	390
Northeast U.S.	1944	3	390
Louisiana (Grand Isle)	1909	4	350
Louisiana (New Orleans)	1915	4	275
Texas (Galveston)	1915	4	275

lightweight structures such as mobile homes and poorly constructed buildings.

Part of wind damage is due to objects that are picked up and hurled through the air. Shingles, aluminum siding, road signs, and any items left outdoors become deadly missiles during a hurricane.

Most wind damage from a hurricane occurs within 125 miles (200 kilometers) of the coast. Once a hurricane travels farther inland, it generally begins to weaken. Occasionally a hurricane will retain its strength for even greater distances. In 1989, for example, Hurricane Hugo ripped through Charlotte, North Carolina, with winds gusting to 100 mph (160 kph). Charlotte is about 175 miles (280 kilometers) inland.

Heavy Rain and Flooding

A hurricane's destruction is certainly not limited to coastal areas. To the contrary, for hundreds of miles inland, and for several days after the hurricane-strength winds have died down, the storm may continue to produce torrential rains and flooding. When an area receives more than 6 inches (15 centimeters) of rain, flooding is likely. Hurricanes typically

Hurricanes

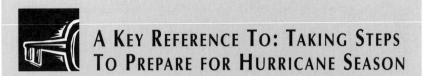

A Key Reference To: Taking Steps To Prepare for Hurricane Season

- Understand the risks that hurricanes pose to your area.
- Learn the routes inland.
- Find out where emergency shelters are located.
- Develop a safety plan for your family.
- Obtain and/or check the batteries in flashlights and battery-powered radios.
- Stock-up on non-perishable foods and bottled water.
- Clean out gutters and down spouts.
- Obtain materials you may need to fortify your home, such as plywood.
- Trim shrubs and trees.
- Make a disaster supply kit. This kit should include (quantities given are per person): non-perishable food; three gallons of bottled water; one change of clothing and footwear; one blanket or sleeping bag; first-aid kit; flashlight, radio, and batteries; extra set of car keys; credit card or cash; and diapers for infants.

drop 5 to 10 inches (13 to 25 centimeters) on the land in their path. Some hurricanes have produced more than 25 inches (63 centimeters) of rain in a twenty-four-hour period.

Inland flooding is the most destructive element of some hurricanes. Examples of this include:

- Hurricane Diane in 1955 brought rains and flooding to Pennsylvania, New York, and New England. The flooding caused nearly 200 deaths and $4.2 billion in damage.

- Hurricane Camille in 1969 brought 9.8 inches (25 centimeters) of rain to Virginia's Blue Ridge Mountains. The storm resulted in 150 deaths.

- Tropical storm Claudette in 1979 dumped 45 inches (114 centimeters) of rain outside of Alvin, Texas, resulting in over $600 million in damage.

It should also be noted that heavy rains due to hurricanes are not always harmful. These rains may spell relief for regions with parched soil

A KEY REFERENCE TO: WHAT TO DO WHEN YOU ARE WITHIN A HURRICANE WATCH AREA

- Stay tuned to radio or television reports of the storm's progress.
- Fill your car with gas and get cash, if needed.
- Cover all windows and doors with shutters or plywood.
- Check your supply of non-perishable food and water.
- Gather first-aid materials and medications.
- Bring lawn furniture, garbage cans, garden hoses, and other lightweight items inside.
- If you have a boat, be sure it is properly secured.
- Evacuate if you live in a mobile home or high-rise, on the coastline, on an offshore island, or near a river or flood plain.

and withering crops. In some cases, the value of saved crops in one region is greater than the value of property destroyed by the flooding in another region. In some areas of the world, particularly in the Far East, farmers are dependent on annual hurricane rains for their economic survival.

TORNADOES

Another hazard of hurricanes is tornadoes. About one-quarter of all hurricanes that come on shore in the United States produce tornadoes. A single hurricane, on average, spawns ten tornadoes.

The **thunderstorm**s embedded in the outer regions of a hurricane are the most likely to spawn tornadoes, although tornadoes also form in thunderstorms close to the **eye wall.** The greatest number of tornadoes are produced in the portion of the hurricane that is northeast of the eye.

Theodore Fujita, a tornado specialist from the University of Chicago, is the author of a recent theory that the greatest damage from hurricane-induced tornadoes is caused by small funnels called **spin-up vortices.** These vortices are each 9 to 30 feet (3 to 10 meters) in diameter and last only about ten seconds. The hurricane winds combine with the tornado winds, so the vortices produce winds of around 200 mph (320

Hurricanes

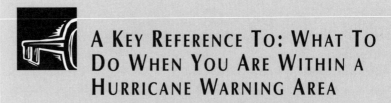

A Key Reference To: What To Do When You Are Within a Hurricane Warning Area

- Stay tuned to radio or television reports of the storm's progress.
- Finish covering your windows and doors and prepare your home for evacuation.
- Evacuate immediately upon the orders of local officials and travel inland, to the home of a friend or relative, a low-rise motel, or an emergency shelter.
- Notify someone outside of the hurricane warning area of your evacuation plans.
- If you have pets that you are unable to take with you, leave them plenty of food and water.

kph). (For more information about Tetsuya Theodore Fujita, see "Thunderstorms," page 238.)

Ranking Hurricanes by Strength

Similar to **tornado**es, **hurricane**s are ranked according to their strength. However, there is one main difference: Tornadoes are classified *after* they have struck an area based on the *actual* damage they have created, while hurricanes are categorized *before* they strike land based on their *potential* damage to coastal areas.

Each hurricane is placed into one of five categories, described in the **Saffir-Simpson Hurricane Damage Potential Scale.** This scale was developed in the early 1970s by Robert Simpson, then-director of the National Hurricane Center, and Herbert Saffir, an engineer who designed Miami's hurricane-proof building code.

According to the Saffir-Simpson Scale, hurricanes in category 1 are the weakest and hurricanes in category 5 are the strongest. The factors that determine a hurricane's strength include: **air pressure** at the **eye** of the storm; range of wind speeds; potential height of the **storm surge**; and the potential damage caused. The categories of potential damage are de-

A Key Reference To: Staying at Home During a Hurricane

- Fill the bathtub and containers with drinking water; unplug small appliances; turn off propane tanks; and turn your refrigerator to its coldest possible setting.
- In case of strong winds: close all outside and inside doors and go into a small interior room or hallway on the first floor, away from windows and doors. If possible, crouch beneath a sturdy piece of furniture.

fined as follows: 1) minimal; 2) moderate; 3) extensive; 4) extreme; and 5) catastrophic. A hurricane's ranking is upgraded or downgraded as it goes through its stages of development.

Saffir-Simpson Hurricane Intensity Scale

Scale number (category)	Central pressure		Wind speed		Storm surge		Damage
	mb	in.	mi/hr	km/hr	ft	m	
1	≥980	≥28.94	74–95	119–154	4–5	1–2	Minimal
2	965–979	28.50–28.91	96–110	155–178	6–8	2–3	Moderate
3	945–964	27.91–28.47	111–130	179–210	9–12	3–4	Extensive
4	920–944	27.17–27.88	131–155	211–250	13–18	4–6	Extreme
5	<920	<27.17	>155	>250	>18	>6	Catastrophic

Category 5 hurricanes are quite rare in the United States. The United States has been struck by only three category 5 hurricanes in this century. These include: a hurricane on Labor Day, 1935 (the practice of hurricane naming did not begin until 1953); Hurricane Camille in 1969; and Hurricane Allen in 1980. On average, only 2 hurricanes of category 3 or greater strike the United States every three years. Just 25 of the 126 tropical storms or hurricanes that hit the United States between 1949 and 1990, or 19.8 percent, were category 3 or higher. Those 25 storms, however, caused 76 percent of all property damage attributed to tropical storms or hurricanes during that time period.

Hurricanes

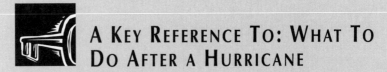

A Key Reference To: What To Do After a Hurricane

- Stay tuned to the radio or television for information.
- Do not return home until your area has been declared safe.
- Do not attempt to drive around a barricade; if you encounter one, turn around and take a different route.
- Do not drive on roads or bridges in flooded areas or on washed-out roads.
- Inspect your gas, water, and electrical lines for damage before using.
- Be sure that your tap water is not contaminated before drinking or cooking.
- Make as few calls as possible so you do not tie up phone lines.

HURRICANE WATCHES AND WARNINGS

The ability of weather forecasters to warn the public of potential **hurricane** strikes has greatly increased in recent decades. As described in the "Forecasting" chapter, some of the sophisticated technologies that are now used to detect and track tropical cyclones include **radar, weather satellites,** and **weather aircraft.** Although the erratic nature of hurricanes makes it impossible to predict exactly when and where they will strike, forecasters can now guess at the answers to these questions with a fair degree of certainty.

When a hurricane threatens a coastal area, hurricane **watch**es and **warning**s are issued. If the hurricane is predicted to be life-threatening, the residents of that area are evacuated. Hurricane watches and warnings are issued by the Tropical Prediction Center (formerly the National Hurricane Center), a branch of the National Weather Service.

A hurricane watch is issued when a hurricane is headed in the general direction of an area. It means that hurricane conditions are *possible* in the area. Hurricane watches are announced at least thirty-six hours, and sometimes several days, in advance.

If a hurricane is poised to strike an area within twenty-four hours, a hurricane warning is issued. It means that hurricane conditions are *ex-*

pected in the area. For each community within the warning area, meteorologists will also issue the probability of the hurricane's center coming within 65 miles (105 kilometers) of that community. The probability is intended to give residents an idea of the type of damage likely to occur in their area. It is also considered in the decision as to whether or not evacuation is necessary.

Typically, an area 340 miles (550 kilometers) long is included in a hurricane warning. This distance is about three times the area that will actually be affected once a hurricane comes onshore. The reason the warning area is so large is that it's possible for a hurricane to change course at any time, making it impossible to predict the exact point at which the hurricane will make **landfall.**

For people who live in areas affected by hurricanes, it is crucial to know and follow certain safety procedures. These procedures include making preparations before the hurricane season begins; knowing what to do when a hurricane is in progress; and taking certain steps after the hurricane has passed.

For more information about the dangers hurricanes pose to your area and hurricane preparedness, contact your local office of the National Weather Service, American Red Cross, or Federal Emergency Management Agency.

NAMING HURRICANES

Before 1950, **hurricane**s were identified primarily by their **latitude** and **longitude.** This practice became confusing as hurricanes moved about, especially if there was more than one hurricane at the same time on the same ocean.

In 1950, meteorologists began the practice of assigning names to all hurricanes and **tropical storms** that formed in the western north Atlantic, Caribbean, and Gulf of Mexico. They began naming eastern Pacific storms in 1959. From 1950–53, names were taken from the international radio codes words that corresponded with letters of the alphabet. For instance, the first three letters—"a," "b," and "c"—had the names Able, Baker, and Charlie.

In 1953, meteorologists began giving the storms female names. The names were assigned in alphabetical order, starting with the "As" for each new season. Since 1978 in the eastern Pacific, and 1979 in the north Atlantic, male names, as well as names in French and Spanish, have also been used.

Hurricanes

Names are now assigned in advance for six-year cycles. The names are submitted by countries that lie in the path of hurricanes and must be approved by the Region 4 Hurricane Committee of the World Meteorological Organization, which is made up of representatives of countries affected by hurricanes.

After the six-year cycle has ended, the names may be re-used. The names of hurricanes that cause extensive damage, however, such as Gilbert, Gloria, Hugo, and Andrew, are removed from the list for at least ten years.

Presently, each hurricane-producing region of the world (except the north Indian Ocean, where cyclones are not named) has its own lists of names, drawn up years in advance. Each storm is automatically assigned the next name on the alphabetical list.

Names for Hurricanes Through the Year 2000

Eastern Pacific Hurricane Names					North Atlantic Hurricane Names				
1996	1997	1998	1999	2000	1996	1997	1998	1999	2000
Alma	Andres	Agatha	Adrian	Aletta	Arthur	Ana	Alex	Arlene	Alberto
Boris	Blanca	Blas	Beatriz	Bud	Bertha	Bill	Bonnie	Bret	Beryl
Christina	Carlos	Celia	Calvin	Carlotta	Cesar	Claudette	Charley	Cindy	Chris
Douglas	Delores	Darby	Dora	Daniel	Dolly	Danny	Danielle	Dennis	Debby
Elida	Enrique	Estelle	Eugene	Emilia	Edouard	Erika	Earl	Emily	Ernesto
Fausto	Felicia	Frank	Fernanda	Fabio	Fran	Fabian	Frances	Floyd	Florence
Genevieve	Guillermo	Georgette	Greg	Gilma	Gustav	Grace	Georges	Gert	Gordon
Hernan	Hilda	Howard	Hilary	Hector	Hortense	Henri	Hermine	Harvey	Helene
Iselle	Ignacio	Isis	Irwin	Ileana	Isidore	Isabel	Ivan	Irene	Isaac
Julio	Jimena	Javier	Jova	John	Josephine	Juan	Jeanne	Jose	Joyce
Kenna	Kelvin	Kay	Kenneth	Kristy	Kyle	Kate	Karl	Katrina	Keith
Lowell	Linda	Lester	Lidia	Lane	Lili	Larry	Lisa	Lenny	Leslie
Marie	Marty	Madeline	Max	Miriam	Marco	Mindy	Mitch	Maria	Michael
Norbert	Nora	Newton	Norma	Norman	Nana	Nicholas	Nicole	Nate	Nadine
Odile	Olaf	Orlene	Otis	Olicia	Omar	Odette	Otto	Ophelia	Oscar
Polo	Pauline	Paine	Pilar	Paul	Paloma	Peter	Paula	Philippe	Patty
Rachel	Rick	Roslyn	Ramon	Rosa	Rene	Rose	Richard	Rita	Rafael
Simon	Sandra	Seymour	Selma	Sergio	Sally	Sam	Shary	Stan	Sandy
Trudy	Terry	Tina	Todd	Tara	Teddy	Teresa	Tomas	Tammy	Tony
Vance	Vivian	Virgil	Veronica	Vicente	Vicky	Victor	Virginie	Vince	Valerie
Winnie	Waldo	Winifred	Wiley	Willa	Wilfred	Wanda	Walter	Wilma	William
Xavier	Xina	Xavier	Xina	Xavier					
Yolanda	York	Yolanda	York	Yolanda					
Zeke	Zelda	Zeke	Zelda	Zeke					

9

TEMPERATURE EXTREMES, FLOODS, AND DROUGHTS

The material contained in this chapter draws on information presented in a number of other chapters. For instance, **air mass**es; global-scale, upper-level winds; and the formation of clouds and precipitation were explained in the chapter "What Is Weather?" **Flash flood**s were described in the chapter "Thunderstorms." Flooding due to **storm surge**s was discussed in the chapter "Hurricanes." And the forces that result in certain regions of Earth being typically or seasonally hot, cold, wet, or dry, are covered in the chapter "Climate."

In this chapter, we will study the circumstances that give rise to extreme weather conditions, describe specific examples of weather extremes around the world, and explore how extreme weather affects humans.

TEMPERATURE EXTREMES

Some parts of the world regularly experience extreme heat or extreme cold. Those locations, in other words, have hot or cold **climate**s. As described in the chapter "Climate" (see page 451), a location's climate depends primarily on its **latitude,** topographical features, altitude, proximity to oceans, proximity to **semipermanent highs** or **lows,** and wind circulation patterns.

Antarctica, for instance, has a bitterly cold climate, while the Sahara Desert has a very hot climate. The world's highest and lowest temperatures are found in those regions that are typically hot and cold, respectively.

However, as we will see, temperature extremes can also be experienced in areas with **temperate,** or mild, climates—areas that occupy the

Temperature Extremes, Floods, and Droughts

middle latitudes. In the middle latitudes, temperatures are considered to be extreme whey they deviate substantially from the norm. Thus, for instance, while 20°F (-7°C) in Antarctica would be considered mild, that same temperature in Florida would be considered extremely cold.

EXTREME TEMPERATURES IN THE MIDDLE LATITUDES

A location's temperature at any given time is determined by the presiding air mass. Tropical air masses bring warmer temperatures and polar air masses bring colder temperatures to the middle latitudes (see "What Is Weather?" on page 18).

One way in which unusually warm or cold temperatures are produced in middle latitudes is that air masses travel farther towards the equator or the poles than usual. For instance, an air mass originating in the arctic may be brought far southward (in the **Northern Hemisphere**) by a deep **trough** within the upper-level, global winds called **upper-air westerlies** (see "What Is Weather," on page 26). The result would be that a southern location, such as Texas, which normally has hot summers and mild winters, would experience very cold air. In contrast, a warm air mass originating near the equator may be brought a great distance northward, in a **ridge** within the upper-air westerlies. As a result, a typically cool place, like northern Minnesota, would experience uncommonly high temperatures.

Figure 37: The formation of blocking systems.

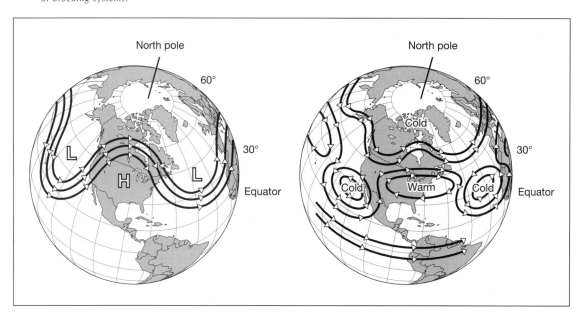

292   THE COMPLETE WEATHER RESOURCE

Blocking Systems

The greatest temperature extremes in the middle latitudes are produced beneath **blocking systems.** A blocking system is a whirling air mass, containing either a high-pressure system (a **blocking high**) or a low-pressure system (a **blocking low**), that gets cut off from the main flow of upper-air westerlies (see Figure 37). A blocking system forms when upper-air westerlies exhibit a strong northward-and-southward wavelike motion. In the presence of a blocking system, the upper-air westerlies are diverted around the blocking system, either to the north or the south.

A blocking system will remain stationary for an extended period of time, sometimes up to several weeks. During that time, the surface air beneath the blocking system may grow exceedingly cold (in winter), or hot (in summer). This effect is caused when the blocking system is cut off from Earth's natural system of heat distribution around the planet. Normally, upper-air winds serve to moderate temperatures by circulating alternating warm and cold air masses.

An example of extreme heat produced by a blocking high occurred during the summer of 1980. The high that produced this effect was located over the southeastern United States. In the southern Mississippi Valley, temperatures rose to nearly 100°F (38°C) every day for over a month. More than 1,250 people died as a result of the heat.

Blocking highs and lows may also lead to **drought**s or **flood**s. A blocking high produces clear weather because air descends beneath it, to the region of lower pressure at the surface. And when air descends, the moisture within it evaporates. If a blocking high remains in place long enough, it can produce a drought. In contrast, a blocking low produces continually cloudy and rainy skies because air rises beneath a low-pressure system, causing **condensation** and cloud-formation to occur. Some of the worst floods in the history of the United States have occurred beneath blocking lows.

Extreme Heat

Extremely hot conditions have been recorded in many parts of the world. Although the tropical deserts are consistently the world's hottest places, temperatures similar to those in tropical deserts are periodically observed in the **middle latitudes,** as well.

One reason that the hottest places are in deserts is that the dryness intensifies the heat. First of all, deserts have few, if any, clouds to block

Temperature Extremes, Floods, and Droughts

incoming solar energy. And air above dry ground heats up faster than air above wet ground. When the ground is wet, some of the sun's heat is absorbed in the process of **evaporation.** When the ground is dry, all incoming solar radiation heats the surface and is transferred, by **conduction,** into the air.

WORLD HEAT RECORDS

When discussing record high and low temperatures, it's important to note that these temperatures are merely *recorded* extremes. In most parts of the world, temperature records have been kept for fewer than two hundred years. That period of time represents only a tiny fraction of the history of Earth. Furthermore, there are still many places on Earth where, usually because of their inaccessibility, weather records are still not kept. Therefore, it is reasonable to assume that there have been times or places where the temperatures have occurred that are higher or lower than those we consider to be "record" highs and lows.

Bearing that in mind, the hottest place in the world is Dallol, Ethiopia, situated at **latitude** twelve degrees North, along the eastern perimeter of the Sahara Desert, near the Red Sea. According to records kept during the years 1960–66, the average annual temperature in Dallol was 93.9°F (34.4°C). Average daily temperatures exceeded 100°F (38°C) for every month of the year except December and January. The average daily temperature for December was 98°F (37°C) while the average daily temperature for January was 97°F (36°C). On many days the temperature rose to above 120°F (49°C).

However, the place with the world's highest recorded temperature is not Dallol. That distinction belongs to El Azizia, Libya, also located on the edge of the Sahara, about 2,170 miles (3,500 kilometers) northeast of Dallol. The record high of 136.4°F (58°C), was recorded in El Azizia on September 22, 1922.

The hottest place in the United States is Death Valley, California. A record high temperature for North America, of 134°F (57°C), was recorded in Death Valley on July 10, 1913. The temperature there reached 127°F (53°C) twice in the 1980s, as well.

Death Valley is consistently hot during the summer months, although it cools significantly in the fall and winter. In July, its hottest month, the average maximum tem-

This auto mechanic works under the slight breeze of a fan during a heat wave in New York City.

> ### EXTREME WEATHER: RECORD HIGH TEMPERATURES AROUND THE WORLD
>
> - Africa (& the world): 136°F (58°C) in El Azizia, Libya on September 13, 1922
> - North America: 134°F (57°C) in Death Valley, California, on July 10, 1913
> - Middle East (& Asia): 129°F (54°C) in Tirat Tsvi, Israel, on June 21, 1942
> - Australia: 128°F (53°C) in Cloncurry, Queensland, on January 16, 1889
> - Europe: 122°F (50°C) in Seville, Spain, on August 4, 1881
> - South America: 120°F (49°C) in Rivadavia, Argentina on December 11, 1905
> - Canada: 113°F (45°C) in Midale, Saskatchewan, on July 5, 1937
> - Great Britain: 100°F (38°C) in Tonbridge, Kent, on July 22, 1868
> - Hawaii: 100°F in Pahala on April 27, 1931
> - Alaska: 100°F in Fort Yukon on June 27, 1915
> - Greenland: 86°F (30°C) in Fvitgut on June 23, 1915
> - Antarctica: 58°F (14°C) in Esparanza on October 20, 1956

perature is 116°F (47°C). Nighttime low temperatures in the summer in Death Valley sometimes remain over 100°F (38°C). Death Valley typically has between 140 and 160 days per year with temperatures over 100°F.

The U.S. city with the highest summertime temperatures is Yuma, Arizona, just across the California state line. In July, Yuma has an average high temperature of 107°F (42°C). There was a period of 101 straight days in 1937 during which the high temperature in Yuma was over 100°F (38°C). The continental U.S. city with the highest year-round average temperature is Key West, Florida, at 78°F (26°C).

During periodic heat waves, extremely high temperatures have been recorded throughout most of the continental United States. Eight

Temperature Extremes, Floods, and Droughts

Weather Report: Heat Bursts

A **heat burst**, also called a warm wake, is a sudden, short-lived, dramatic warming of the air that is produced in the wake of a dissipating **thunderstorm.** Specifically, a heat burst is a warm, dry **downdraft,** a downward gust of air (see "Thunderstorms," page 215). The downdrafts from thunderstorms are usually cool. However, when the downdraft originates high above the surface, say between 10,000 and 20,000 feet (6,000 and 12,000 meters), its temperature raises considerably by **compressional warming** and it reaches the ground as a warm gust of wind.

Compressional warming, as we learned in "What Is Weather," occurs as an air parcel descends and is compressed by the increasing pressure of the surrounding air. That compression leads to a greater number of collisions between molecules, which causes an increase in the temperature of the air.

Heat bursts are rare. When they do occur, it is usually during the summer and at night. It is possible, however, that heat bursts occur with equal frequency during the day and at night, but are just much more noticeable at night. That is because at night, when temperatures are lower than they are during the day, a heat burst produces a much more dramatic warming.

A heat burst on September 9, 1994, produced a warming sufficient to tie the record high temperature in Glasgow, Montana. According to radio station KFBB in nearby Great Falls, Glasgow's temperature rose from 67°F (19°C) at 5:02 A.M., to 93°F (34°C) at 5:17 A.M. By 5:40 A.M., however, the temperature was back down to 68°F (20°C).

A heat burst produced record heat in Portugal on July 6, 1949. According to a meteorological observer there, the temperature climbed from 100°F (38°C) to 158°F (70°C) in just two minutes. However, this report is not considered the official world heat record since independent verification was not possible.

states in the United States have record highs of 120°F (49°C) or greater. They are: California with 134°F (57°C); Arizona with 128°F (53°C); Nevada with 125°F (52°C); North Dakota and Kansas with 121°F (49°C); and Oklahoma, South Dakota, and Texas with 120°F (49°C). In El Paso, Texas, the daily high temperature exceeded 110°F (43°C) for an entire week in June, 1994.

HEALTH RISKS OF EXTREME HEAT

Hot weather adversely affects human health in a number of ways, stemming from the raising of the body temperature. Raising the body temperature, in turn, can lead to dehydration, **heat exhaustion,** and **heat stroke.** Heat, especially in combination with humidity, places added stresses on the body's circulatory system. For elderly people or people with serious illnesses, that added stress of heat can prove fatal. The most common causes of death in hot weather are cardiac arrest, stroke, and respiratory distress.

Many people each year perish from the heat. **Heat wave**s, extended periods of high heat and humidity, claim approximately 175 to 200 lives each summer in the United States and a far greater number in developing nations. The number of deaths per day increases the longer a heat wave persists.

The human body attempts to dissipate heat in two ways: through perspiration and by varying the rate of blood circulation. Panting is a third cooling mechanism that becomes a factor only when the blood is heated above 98.6°F (37°C).

Perspiration fulfills 90 percent of the body's cooling function. Sweat is exuded from more than 3 million sweat glands covering the entire body. When the air is relatively dry, sweat evaporates quickly. This evaporation removes heat from the skin, thus cooling the body (see "What Is Weather?" box on page 47).

Perspiration also has a negative effect: it takes water out of the body. An adult walking in the hot desert can lose more than 3.5 pints (1.7 liters) of water through perspiration in just one hour! If you walk in Death Valley without drinking water, you can become severely dehydrated within minutes. And once the body becomes dehydrated, its cooling mechanism becomes crippled.

Heat is even more dangerous to human health in humid conditions than it is in dry conditions. Sweat evaporates much more slowly into

Temperature Extremes, Floods, and Droughts

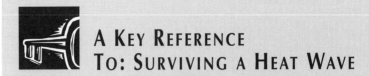

A Key Reference To: Surviving a Heat Wave

- Avoid overexertion, especially during the hottest part of the day. The coolest part of day, and the safest time for strenuous activity, is between 4 A.M. and 7 A.M.
- Wear loose, lightweight, light-colored clothing. This type of clothing allows air to circulate while protecting the skin from the sun's heat and damaging rays. The light color reflects, rather than absorbs, sunlight.
- Remain indoors as much as possible. If you don't have air conditioning, stay on the first floor, out of the sunshine, and keep the air circulating with electric fans. Each day that the air is very hot, try to spend some time in an air-conditioned environment.
- Drink plenty of fluids, even if you don't feel thirsty. Water is best. Avoid drinks with caffeine or alcohol, since those drinks dehydrate the body.
- Eat frequent, small meals.
- Avoid overexposure to the sun. When skin is sunburned, its ability to dissipate heat is hampered.
- If you are taking medication that affects your blood circulation, ask your physician how that medication affects your ability to tolerate heat.
- Groups of people most susceptible to heat-related illness include: elderly people, small children, people with chronic illnesses, overweight people, and people with alcohol dependency. People in those groups should be especially cautious during a heat wave.

humid air than it does into dry air. Thus, the body cools much more slowly.

Another response to an increase in body temperature is that the heart pumps faster. Blood vessels become dilated and blood is directed into tiny capillaries near the surface of the skin, so that it can cool.

People undertaking physical activity in hot, humid conditions, particularly if they're not drinking enough, may fall victim to one or more heat-related illnesses. The onset of heat-related illness begins when one's body temperature rises more quickly than the body's cooling mechanisms can dissipate the heat.

Heat-related illnesses range from relatively harmless **heat cramp**s to potentially fatal heat stroke. The severity of the illness generally increases with age. Under similar conditions, a young adult may experience only mild heat cramps whereas an elderly person may experience severe heat stroke.

Heat cramps are muscle cramps or spasms, usually afflicting the abdomen or legs, caused by exercising in hot weather. They are caused by a temporary imbalance in body salts that may occur during heavy perspiration. Heat cramps mostly afflict people who are used to cooler weather and occur less frequently as one becomes acclimatized to the heat.

The remedy for heat cramps is to move to a cool place, rest, and drink fluids. Light stretching of the affected muscles is also recommended.

Another possible consequence of exercising in hot weather is **heat syncope,** or fainting. This condition results from a rapid drop in blood pressure. Most people recover quickly from heat syncope by resting in a cool place and drinking fluids.

A more serious condition is heat exhaustion. Heat exhaustion, a form of mild shock, results from fluid and salt loss through heavy perspiration. In addition, when a greater proportion of blood flow is toward the skin, less blood reaches the vital organs.

The symptoms of heat exhaustion include flushed skin, a slight fever (not higher than 102°F, or 39°C), weakness, and a headache. There may also be dizziness, nausea, and vomiting. A person with heat exhaustion feels hot and thirsty and his or her skin becomes sweaty.

A person who falls victim to heat exhaustion should be moved to a cool place, their clothing removed or loosened, and cool, wet towels applied. He or she should be encouraged to drink cool water, slowly. The recommended amount is one-half glass every fifteen minutes. Hospitalization may be required for elderly people who experience heat exhaustion.

Heat stroke (also called "sunstroke") is a life-threatening condition that develops when heat exhaustion is left untreated. Heat stroke begins when the body has exhausted its efforts to cool itself.

If someone has heat stroke, his or her skin will be hot and red and his or her breathing will be rapid and shallow. The victim will cease to sweat and his or her temperature rapidly rises to 105°F (40.5°C) or even

Temperature Extremes, Floods, and Droughts

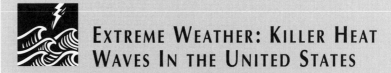

EXTREME WEATHER: KILLER HEAT WAVES IN THE UNITED STATES

- During the hot, dry "dust bowl" summer of 1936, when high temperature records were established in many places around the United States, about 15,000 people died.
- Over 9,500 people perished in the heat wave of the summer of 1901.
- The Midwest heat wave of 1980 claimed over 1,250 lives.
- Up to 10,000 people died in the heat wave and drought of 1988.
- More than 700 people in Chicago, most of them elderly, died from the heat during the heat wave of July 12–15, 1995. The temperature during that period peaked at 104°F (40°C). When combined with the humidity, it felt more like 119°F (48°C).

higher. The person will no longer feel hot or thirsty and may become confused and delirious, experience seizures, and lose consciousness.

A person with heat stroke requires emergency medical attention. Call 911 and attempt to cool the person. Place the person in a cool bath or wrap him or her in cool, wet sheets. If the person is conscious, offer him or her small amounts of cool water. If the person does not want to drink, do not force him or her to do so.

If left untreated, heat stroke can lead to brain damage and failure of vital organs. It is rare for someone to survive once their body temperature exceeds 108°F (42°C).

Heat also has indirect health consequences. Hayfever and asthma attacks, for instance are triggered by pollen and other allergens spread by rising air currents. In addition, bacteria thrive in warm weather. And as a consequence, diseases spread more rapidly.

Devastating disease outbreaks often occur during heat waves where conditions are unsanitary, particularly in developing nations. One of the greatest health problems in hot, wet weather, is malaria outbreaks. These outbreaks occur because malaria-carrying mosquitoes hatch in warm, wet conditions.

An increase in violent behavior also seems to coincide with heat waves. During the heat wave of 1988, for instance, the murder rate in New York City increased 75 percent. While it's possible that heat triggers violence, an alternative theory is that people consume more alcohol in hot weather and that drunkenness leads to violence.

EXTREME COLD

The consistently coldest places in the world are near the poles, the South Pole in particular. While the **middle latitudes** sometimes experience bitter cold weather, those temperatures never approach the extremely low temperatures found near the poles.

The extreme cold of the polar regions is brought about, primarily, by two factors. The first is limited solar radiation. Due to the tilt of Earth's axis, the poles receive only very indirect sunlight. And for the six months of the year centered around the winter solstice (the shortest day of the year), the sun either stays below the horizon or barely rises above it (see "Climate," page 476). For the other six months of the year, the sun never gets very high above the horizon.

The second factor is that the land and frozen seas around the poles are covered with snow. Snow has a reflectivity of about 90 percent. As a result, most of what little sunlight does reach the poles is reflected back into the atmosphere, leaving only a small portion to be absorbed by the ground.

WORLD COLD RECORDS

The coldest place in the world is Antarctica. Temperatures vary greatly throughout the continent, with the warmest temperatures found on the coasts and the coldest temperatures found in the interior. In the summer months (October through March), average temperatures range from 14 to 21.2°F (-10 to -6°C) near the seas and from -22 to -4°F (-30 to -20°C) near the center of the continent. In the coldest months (April through September), the outer areas have average temperatures around -22°F (-30°C), while the average temperature in the interior is around -94°F (-70°C).

Earth's lowest temperature, -128.6°F (-89.2°C), was recorded on July 21, 1983, at the Soviet research station Vostok in Antarctica. This station is located at **latitude** of 78 degrees south and an elevation of 11,441 feet (3,488 meters) above sea level. During Vostok's warmest month, December, the average temperature is -27°F (-33°C) and during

Temperature Extremes, Floods, and Droughts

its coldest month, August, the average temperature is -91°F (-68°C). The mean annual temperature at Vostok is -72°F (-58°C). The warmest temperature ever recorded at Vostok was -5.8°F (-21°C).

The mean annual temperature at the geographic South Pole is -56°F (-49°C). These temperatures have been recorded since the late 1960s by a U.S. outpost at the geographic pole, nearly 9,200 feet (2,800 meters) above sea level. The mean temperature in July, the coldest month, is -75°F (-59°C). The record low temperature at the South Pole is -117°F (-83°C), recorded on June 23, 1983.

The north polar region, particularly the northern reaches of Greenland and Siberia, also experiences extreme cold temperatures, although not to the same degree as its southern counterpart. Average temperatures near the North Pole are about -43°F (-42°C) in January, the coldest month, and about 8.6°F (-13°C) in July, the warmest month.

The two coldest temperatures ever recorded in the north polar region were both found in Siberia. The temperature dropped to -96°F (-71°C) in Oymyakon, at a latitude of 63 degrees North, in 1964. However, since the reading was taken on a **thermometer** that was not of the type used by world meteorological agencies, this reading is not considered an official record. The official record was set when the mercury fell to -90°F (-68°C) in Verkhoyansk, at a latitude of 67 degrees North, in 1892. The average January temperature in Yakutsk, Siberia, at a latitude of 62 degrees North, is -46°F (-43°C), and the average yearly temperature is 12°F (-11°C).

The coldest temperature ever recorded in Greenland was -87°F (-66°C) on January 9, 1954, at Northice, which is at a latitude of 78 degrees North, and an altitude of 7,685 feet (2,343 meters). Farther south, at a latitude of 71 degrees North, is the town of Eismitte (which means "middle of the ice"). At this location, 9,941 feet (3,031 meters) above sea level, the average temperature in February, the coldest month, is -53°F (-47°C) and the average temperature in July, the warmest month, is 12°F (-11°C). The average yearly temperature for Eismitte is -22°F (-30°C).

The coldest place in the continental United States is International Falls, Minnesota, at a latitude of 49 degrees North, on the Canadian border. The average January temperature there is 1°F (-17°C). Minnesota is also the state with the coldest urban area in the United States. Minneapolis/St. Paul, located 250 miles (400 kilometers) south of International Falls, has an average temperature of 16°F (-9°C) during its coldest months. Another one of the coldest cites in the lower forty-eight states is Butte, Montana. The temperature there dips below the freezing point for 223 days per year, on average.

The city within the continental United States that holds the record for the longest cold period is Langdon, North Dakota. There the temperature remained below 0°F (-18°C) for forty-one straight days from January 11 to February 20, 1936. And temperatures remained below 32°F (0°C) for ninety-two straight days in Langdon from November 30, 1935, to February 29, 1936.

The record for the lowest temperature in the continental United States is held by Rogers Pass, Montana, located at a latitude of 47 degrees North. On January 20, 1954, the temperature was -70°F (-57°C). And the record cold for Alaska is -80°F (-62°C), taken on January 23, 1971, in Prospect Creek, in the Endicott Mountains, at a latitude of 66 degrees North.

Sixteen states in the United States have record lows of -50°F (-45°C) or lower. They are: Alaska with -80°F (-62°C); Montana with -70°F (-57°C); Wyoming with -66°F (-54°C); Colorado, with -61°F (-52°C); Idaho and North Dakota with -60°F (-51°C); Minnesota with -59°F (-51°C); South Dakota with -58°F (-50°C); Oregon and Wisconsin with -54°F (-48°C); New York with -52°F (-47°C); Michigan with -51°F (-46°C); and Nevada, New Mexico, Utah, and Vermont with -50°F (-46°C).

The coldest parts of North America are the Northwest Territories and Yukon of Canada. Resolute, which is on Cornwallis Island, Canada, in the Arctic Ocean at a latitude of 75 degrees North, has an average January temperature of -26°F (-32°C).

HEALTH RISKS OF EXTREME COLD

Cold weather has a number of adverse effects on the human body. Some of these are direct results of the cold, such as **frostbite,** which is the freezing of the skin, and **hypothermia,** which is the drop in core body temperature from the normal 98.6°F (37°C) down to 95°F (35°C) or less. Other maladies are indirect results of the cold, and accompanying snow and ice, such as traffic accidents and heart attacks suffered while shoveling snow.

The human body has little natural protection against the cold. It has far less ability to adapt to cold weather than it does to hot weather. Without the proper clothing in cold weather, a person rapidly loses body heat.

Temperature Extremes, Floods, and Droughts

EXTREME WEATHER: RECORD LOW TEMPERATURES AROUND THE WORLD

- Antarctica (& the world): -128.6°F (-89.2°C) in Vostok on July 21, 1983
- Russia (& the Northern Hemisphere): -90°F (-68°C) in Verkhoyansk, Siberia, on February 5 and 7, 1892
- Greenland: -87°F (-66°C) in Northice on January 9, 1954
- Canada (& North America): -81°F (-63°C) in Snag, Yukon, on February 3, 1947
- United States: -80°F (-62°C) in Prospect Creek, Alaska, on January 23, 1971
- Continental United States: -70°F (-57°C) in Rogers Pass, Montana, on January 20, 1954
- South America: -27°F (-33°C) in Sarmiento, Argentina, on June 1, 1907
- United Kingdom: -17°F (-27°C) in Braemar, Scotland, on February 11, 1985, and January 10, 1982
- Greece: -13°F (-25°C) in Kavala on January 27, 1954
- Africa: -11°F (-24°C) in Ifrane, Morocco, on February 11, 1935
- Australia: -8°F (-22°C) in Charlotte Pass on July 22, 1949
- Hawaii: 14°F (-10°C) in Mt. Haleakala on January 2, 1961

Even at temperatures as high as 68°F (20°C), an unclothed person will begin to shiver. Children and elderly people are the least able to withstand cold weather since their bodies are less efficient at regulating temperature than people in other age groups.

When combined with wind, cold weather is even more hazardous. The **windchill equivalent temperature** is the *apparent* temperature, or how cold it feels, when the wind speed is factored in with the temperature. For instance, when it is 0°F (-18°C) outside and the wind is blowing at 20 mph (32 kph), it feels like -40°F (-40°C) (see "What Is Weather?" on page 9). There is a risk of flesh freezing when the windchill is below -22°F (-30°C) and it takes only a minute or so for flesh to freeze when the windchill is below -58°C (-50°C).

Researchers stationed in Antarctica follow a "30-30-30 rule." That rule states that with a temperature of -30°F (-34°C) and wind at 30 mph (48 kph), a person can only survive for 30 minutes.

If the body temperature lowers to the point that hypothermia sets in, a person will experience a gradual decline in physical and mental functions. At the onset of hypothermia, a person will shiver violently. As the body temperature continues to drop, however, the shivering will decrease. Advanced hypothermia can produce unconsciousness and even death.

Another consequence of exposure to extreme cold is frostbite. Frostbite is the freezing of the skin, which causes tissue damage. Frostbite is common in those who engage in winter sports. Cases of frostbite range from mild to severe. Complete recovery is possible in the mildest cases. It is normal for the frostbitten area to feel somewhat numb for several months after frostbite. Serious cases of frostbite can produce a long-term sensitivity to the cold. In the most severe cases, when the tissue freezes to the point that it dies, the affected area will turn black. In such cases it is sometimes necessary to amputate the affected area. The most susceptible parts of the body to frostbite are the ears, nose, hands, and feet.

The earliest warning that you are in danger of getting frostbite is that your fingers, toes, or nose begin to hurt. Frostbite is actually beginning if any of those body parts starts to feel numb. If this occurs, get out of the cold immediately. Warm up the frostbitten area by rubbing it gen-

Humans have little natural protection against extreme cold.

Temperature Extremes, Floods, and Droughts

tly. If it hurts as it gets warmer, this is a good sign. It means that the tissues are still alive.

Cold weather is also hard on humans due to a host of secondary factors. There is a greater incidence of circulatory, respiratory, and infectious diseases during cold weather. Greater stress is placed on the heart during cold weather, since the outer parts of the body become cool. The heart compensates for this by working harder to pump blood to those areas.

The cooling of the body also reduces people's resistance to viral and bacterial infection. That fact, combined with the tendency of people to spend more time indoors in the winter, where they are exposed to others' germs, results in more illness in winter than any other time of year. In countries that experience the range of warm and cold seasons, mortality rates are highest in winter.

Strenuous activity in the winter, such as shoveling snow, can prove fatal to people with circulatory problems. Travel is also more hazardous in the winter. During the eastern United States' cold winter of 1993–94, at least 113 people died from car accidents, heart attacks while shoveling snow, and exposure.

Cold weather also takes a psychological toll. Studies in North America have shown that spending extended periods of time indoors may lead to a condition known as "cabin fever," the symptoms of which are depression, anxiety, and irritability. Another recently named condition, seasonal affective disorder (SAD), is caused by the lack of sunshine during the winter months. A person afflicted with SAD feels depressed and lethargic. A newly developed treatment for SAD is light therapy, in which the patient is exposed to bright lights for certain periods of time each day.

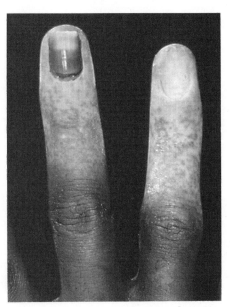

White, speckled skin denotes frostbite, seen here on human fingers.

FLOODS

A **flood** is the inundation of normally dry land with water. Flooding results when the water level, in a river or other body of water, rises and overflows the banks. Floods occur irregularly in some parts of the world and regularly in others. While for some areas flooding spells disaster, for others areas yearly flooding is necessary to sustain crops.

Floods, and **flash flood**s in particular, kill more people than any other weather phenomenon. Worldwide,

>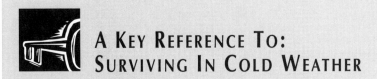
> **A KEY REFERENCE TO: SURVIVING IN COLD WEATHER**
>
> - Be aware of the current temperature and windchill. When dangerous conditions are present, venture out for only short periods.
> - Wear several layers of lightweight clothing.
> - Wear a warm hat with ear flaps, mittens, a covering on your face and neck, warm socks, and waterproof boots.
> - Wear wool clothing closest to your skin. Wool will trap your body heat even if it gets wet.
> - Walk carefully over icy ground.
> - When shoveling, take frequent breaks to avoid overexertion.
> - If you're trapped outdoors in a **blizzard,** dig a large hole in the snow and climb in. This "snow cave," as it is called, will protect you from the wind and decrease the rate at which your body loses heat.

40 percent of all deaths from natural disasters are due to floods. For a list of safety procedures in a flash flood, which are applicable to any type of flood, see "Thunderstorms," pages 240–241.

There are two main types of floods: **coastal flood**s and **river flood**s.

COASTAL FLOODS

Coastal floods are floods that occur along the coasts of lakes and oceans. This type of flooding is of great concern in many countries because of the high population density along their coastlines. In the United States, for example, about 50 percent of the population is concentrated on the coasts.

There are two main causes of coastal flooding: high waters and the **subsidence,** or lowering, of coastal lands. Most coastal flooding is produced by high waters associated with hurricanes. As a hurricane crosses over land, it produces a **storm surge.** A storm surge is a several-foot-high wall of water that results in the flooding of 40 to 100 miles (65 to 160 kilometers) of coastline (see "Hurricanes," page 278).

Temperature Extremes, Floods, and Droughts

Large waves can also cause flooding. The most common type of waves, those driven by the wind, are called **wind waves** (see "Local Winds," page 147). The largest wind waves are generated by large, stationary storm systems. Wind waves tend to reach their greatest heights in the open ocean and diminish in size as they approach land. Large wind waves have the greatest potential for flooding when accompanied by high tide.

The largest waves, however, are generated not by the wind but by submarine earthquakes, landslides, and volcanic eruptions. Waves produced by these forces are called **tsunami**s [tsoo-NAH-meez]. Tsunamis start out small and grow larger as they near land. They travel at speeds of up to 500 mph (800 kph). It is typical for a tsunami to measure 60 to 100 feet (18 to 30 meters) in height by the time it reaches land.

Tsunamis occur most often in the Pacific Ocean. Several tsunamis have affected Alaska and Hawaii. In 1958, a 200-foot-tall (60-meter-tall) tsunami, generated by a minor earthquake and resultant rockfall into the sea, crashed into Lituya Bay, Alaska. Its destroyed great tracts of forest land as far as 1,700 feet (520 meters) above sea level.

Subsidence is the lowering of land in coastal areas, which makes them susceptible to flooding. Subsidence is usually caused by the gradual settling of subterranean rocks or sediments. Subsidence may also be dramatic. For example, if the roof of a cave collapses, the result is a large depression called a **sinkhole.**

River Floods

The banks of rivers and streams overflow due to a host of causes, including excessive rain, the springtime melting of snow, and blockage of water flow due to ice. The failure of a dam or aqueduct is another cause of flooding. The primary cause of flooding in large rivers, such as the Mississippi, the Ohio, and the Missouri, is excessive, prolonged **precipitation** over a large area—sometimes hundreds of square miles. In some areas, flooding occurs nearly every spring due to melting snow.

The most dangerous form of river flood is a flash flood. As we learned in the chapter "Thunderstorms" (see page 238), a flash flood is a sudden, intense, localized flood. It is caused by persistent, torrential rainfall that falls from a slow-moving or stationary severe **thunderstorm.** It may also be caused by the failure of a dam or levee. A dam is a structure that controls the rate of water flow while a levee is a structure built to prevent water from overflowing the banks of a river.

Flash floods occur and recede much more quickly than other types of river floods. They are the most dangerous because they come on so quickly that people are often unable to reach higher ground in advance of the floodwaters. A series of flash floods in the streams that feed into a large river may result in the flooding of the large river.

Flooding may even be caused by a small amount of rain that falls on a blanket of snow. If the ground beneath the snow is frozen, the melting snow and rain will not percolate into the ground, but will run off and drain into a river. The likelihood of flooding is increased if ice blockage prevents water from flowing through the river.

HISTORICALLY SIGNIFICANT FLOODS

- Somewhere between 900,000 and 2.5 million people living in the crowded river valleys perished during the 1887–88 flooding of China's Yellow River. This flood was, by far, the most devastating in recorded history. The Yellow River, nicknamed "China's Sorrow," is one of the world's most flood-prone rivers. Records show that there have been around 1,500 floods on the Yellow River over the last 3,500 years, many of which killed large numbers of people.

- In 1889, a dam break in Johnstown, Pennsylvania, resulted in the worst flash flood in the history of the United States. A wall of water 36 to 40 feet (11 to 12 meters) tall overwhelmed the town, killing over

Temperature Extremes, Floods, and Droughts

Rising flash flood waters in Los Angeles, California, trapped this driver in an intersection, causing him to bail out of his car and jump from hood to hood to reach safety.

Temperature Extremes, Floods, and Droughts

2,200 people. A second flash flood hit Johnstown in 1977, in which 77 people lost their lives.

- The Great Mississippi River Flood of 1927 caused widespread damage over a large area, extending from southern Illinois to the Gulf of Mexico. Flooding began after a year of above-normal precipitation. Water levels along the Mississippi, plus rivers flowing into the Mississippi, were already high when 9 inches (22 centimeters) of rain fell on southern Missouri and most of Arkansas in April.

  Beginning in April, the river flooded over hundreds of square miles. Levees failed in more than 120 places on the Mississippi and its tributaries. Over 26,000 square miles of land across 7 states were flooded. Six-hundred thousand people were forced from their homes and, despite advance warnings, 246 people were killed.

- In the summer of 1993, excessive rains caused what is called "The Great Flood of 1993," one of the greatest floods in the history of the United States. Water overflowed the banks of the Mississippi and Missouri Rivers, inundating 16,000 square miles (41,000 square kilometers) of land in Iowa, Illinois, Minnesota, Missouri, Wisconsin, South Dakota, Nebraska, and Kansas. The Mississippi River swelled to over 7 miles (11 kilometers) wide in some places. Fifty-six small towns along the river were completely submerged in water.

  The flooding was preceded by heavy snowfalls throughout the Midwest in December 1992. The snow melted in March, 16 inches (41 centimeters) of rain fell in the upper Mississippi Valley in April, and the rains continued through July. In mid-June, a **stationary front** formed across the upper Midwest, producing daily thunderstorms.

  The water reached record heights at many points along the Mississippi River and more than 60 percent of levees on that river were destroyed. Forty-eight people died. There was $6.5 billion in crop damage, out of a total property damage of $15-20 billion. Forty-five thousand homes were damaged or destroyed and 85,000 people were evacuated from their homes. Four hundred four counties were declared disaster areas.

- One of the worst floods in the history of the United States ravaged parts of the Midwest and Southeast in March 1997. The disaster was touched off when a series of severe thunderstorms and **tornado**es swept through Arkansas, Tennessee, and Mississippi, killing at least 25 people in Arkansas. Flash floods and gusting winds killed 36 more people in Kentucky, Ohio, West Virginia, and Tennessee, bringing the death toll to 61.

Due to the tornadoes, 11 counties in Arkansas were declared federal disaster areas. And due to floods, 16 counties in Ohio and 63 counties in Kentucky were declared federal disaster areas. Counties that are designated "federal disaster areas" are eligible to receive federal disaster aid.

During the floods, intermittent heavy rains fell throughout the region, accompanied by melting snow. On a single day, March 1, nearly a foot of rain fell on north-central Kentucky, southern Ohio, and west Virginia. New rainfall records were set in Louisville, Kentucky, on that day. Two days later, up to 9 more inches (23 centimeters) of rain fell on the same area.

As a result, flooding occurred along the Ohio, Missouri, and Mississippi rivers, as well as along numerous tributaries. Rivers throughout Kentucky were pushed to record levels. Many river towns in Ohio, Kentucky, Tennessee, and West Virginia were swamped.

Temperature Extremes, Floods, and Droughts

Boating through a flooded street in Macon, Georgia.

Temperature Extremes, Floods, and Droughts

A newspaper account of the flood damage in Falmouth, Kentucky, on March 7, read as follows: "Homes ripped from their foundations and flung 100 feet. Houses tossed into the middle of roads. Trailers packed together. Cars crumpled and even piled atop each other. All coated with an ankle-deep brownish goo."

Thousands of people from flooded river towns were left homeless. The estimated flood damage in Kentucky was $232 million and in Ohio, $40 million.

DROUGHTS

A **drought** is an extended period of abnormal dryness. A drought may last for years and often covers an area the size of several states or more. In a drought, the rate of water loss from soil and plants into the atmosphere greatly exceeds the rate of **precipitation.** Droughts may lead to crop losses, decreased river flow, soil erosion, death of livestock, and even famine. Droughts are experienced throughout the world, although some areas experience droughts much more often than others.

TYPES OF DROUGHTS

There is no uniform set of conditions that defines a drought. A drought is a period of "abnormal" dryness and what is considered "abnor-

The lake at Armour Station, Missouri, vanished due to drought. The baked soil had cracks from 10 to 14 inches (25 to 36 centimeters) deep.

mal" varies between regions with different climates. For instance, in places with a dry climate, such as Australia, a drought is defined as a year in which precipitation is less than 10 percent of average. In contrast, in places with distinct wet and dry seasons, such as India, a drought is defined as a year in which precipitation is less than 75 percent of average.

In the United States, most of which has a **temperate,** humid **climate,** drought is defined as a period at least twenty-one days long during which rainfall over an extensive area is, at most, 30 percent of normal.

CAUSES OF DROUGHTS

The primary cause of droughts in the **middle latitudes** is the prolonged presence of a **blocking high** (see page 293). In addition, droughts are more likely to occur in areas where vegetation has been cleared. The reason for this fact is that plants are important sources of water vapor on land. Water vapor rises and condenses to form clouds and precipitation. The air that rises over ground that has been cleared of plants is drier than the air that rises over ground where plants are present.

A common reason for vegetation loss is overgrazing. Overgrazing results when large numbers of cattle are placed on a tract of land that cannot support them. The cattle eat all the plants, leaving the ground barren. Drought-inducing vegetation loss also occurs at the hands of humans. An example of this phenomena is in the famous "Dust Bowl" (see box, page 315), an event that occurred when farmers cleared grassland to plant wheat, with disastrous effects.

EXAMPLES OF DROUGHTS

- From 1978 to 1980, Australia experienced its worst drought in recorded history. By January 1980, water reservoirs were filled only to 9 percent of capacity and almost all non-irrigated land throughout much of the country was barren. Thousands of kangaroos, plus many sheep and cows, perished. Australia was again beset by drought in the late 1980s. That drought, which lasted into the early 1990s, caused the deaths of 6 million sheep. Many farmers lost as much as 90 percent of their flocks.

- A drought occurred in a ten-state region of the southeastern United States in May through early July of 1986. The states beneath a blocking high, stretching from Virginia and the Carolinas in the east, and southwestward to Mississippi, experienced long periods of clear

Weather Extremes: The Dust Bowl

Temperature Extremes, Floods, and Droughts

The Dust Bowl is the popular name for the approximately 150,000 square miles (400,000 square kilometers) in the South Great Plains region of the United States. This region, which includes the panhandles of Oklahoma and Texas, and the surrounding portions of Colorado, Kansas, and New Mexico, is characterized by low annual rainfall, a shallow layer of topsoil and high winds. The Dust Bowl is best known for the severe drought and violent dust storms it experienced between 1934 and 1937, during the Great Depression.

Pioneers who settled in the Dust Bowl in the 1920s and 1930s made the mistake of plowing up the grassland and planting wheat. As it turned out, the topsoil had been anchored only by the grasses' intricate roots system. Thus when the drought began in 1934, combined with the normally high winds, there was nothing to prevent the disastrous soil erosion that ensued.

The blowing soil covered roads. Drifts in some places were high enough to bury houses. Dust clouds spread over hundreds of miles and were sometimes dense enough to block out the sun. About 300 million tons of topsoil blew away in a single dust storm in May 1934. And on April 14, 1935, a day known as "Black Sunday," a dust cloud in Stratford, Texas, was so thick that some residents suffocated even though they were wearing face masks.

Over 350,000 Dust Bowl farmers, more than half the population of the area, deserted the region. Most of them headed for California. They were nicknamed "Okies" and immortalized in John Steinbeck's classic novel *The Grapes of Wrath*.

After that tragedy, the federal government began a program of land rehabilitation in the Dust Bowl. Grass and trees were planted and appropriate agricultural methods were introduced. These actions eventually made it possible to farm the land again.

The Dust Bowl was beset by other, less severe droughts in the 1950s and 1960s.

Opposite page: A deserted farm in South Dakota, surrounded by drifted dust and sand.

Temperature Extremes, Floods, and Droughts

weather. The states in the center of this region—Kentucky, Tennessee, and North Carolina—were hardest hit by the drought.

- The costliest drought in U.S. history occurred in the summer of 1988. The drought was widespread, affecting over 200 million people in the central and eastern United States, and Canada. The drought reached its peak in July, when it spread across more than 40 percent of the land of the United States.

  It was so hot and dry during July that in parts of the Midwest **lightning** set fire to people's lawns, and the rain from **thunderstorm**s evaporated in mid-air. Between 5,000 and 10,000 deaths were attributed to the heat. Economic losses from the drought in the United States were over $40 billion.

- During the fall of 1995 through the summer of 1996, a severe drought struck the agricultural regions of the southern Great Plains states. Texas and Oklahoma were hardest hit. Over $4 billion dollars in damage resulted from the drought.

10

OPTICAL EFFECTS

The interaction between sunlight and the atmosphere produces a wide array of patterns and colors of light in the sky, and even optical illusions on the ground. The amount and types of particles (including water and ice) in the air, as well as the position of the sun in the sky, influence the quality of the light perceived by our eyes.

Sunlight can appear as white light, as a single color of light, or as the entire spectrum of colors of light. The sun itself may appear as a ringed image or as multiple images. And sunlight can be distorted to create **mirage**s, such as wet roadways or towering mountains on the horizon.

In this chapter we will examine the properties of light and numerous atmospheric optical phenomena that are produced under specific conditions. Most of these phenomena take place during the day although some, caused by the interaction of moonlight and the atmosphere, occur at night. We will also describe the visual displays that take place in the outermost reaches of our atmosphere, the **aurora** borealis and aurora australis—the northern and southern lights.

THE COLOR OF LIGHT

About 45 percent of the solar radiation that reaches Earth's atmosphere is in the form of visible light. Visible light represents the portion of the **electromagnetic spectrum** that we can see. Visible light includes the wavelengths of every color, from that with the longest wavelength, red, to that with the shortest wavelength, violet. The order of wavelengths of colors, from longest to shortest, can be remembered by using the

Optical Effects

mnemonic ROY G. BIV: R=red, O=orange, Y=yellow, G=green, B=blue, I=indigo, and V=violet.

Visible light is broken down into its spectrum of colors when it passes through a glass prism or another medium, such as an ice crystal or raindrop. The prism bends each component of white light to a different degree, depending on that component's wavelength. For instance, red light, which has the longest wavelength, is bent the least. Violet light, which has the shortest wavelength, is bent the most. As a result, the entire rainbow of colors exits the prism, with red and violet on opposite ends.

Sunlight is white, because it contains all visible wavelengths of light. The color of objects is caused by the fact that they absorb some wavelengths of light and reflect others. **Reflection** means that light bounces off a surface at the same angle that it strikes a surface. This definition will become important later in this chapter, when we compare reflection to **refraction,** which is the bending of light.

A white shirt, for example, does not absorb any wavelengths of visible light. The shirt reflects all wavelengths of visible light, which causes it to appear white. On the other hand, the skin of a red apple absorbs all wavelengths of radiation except red. Red light is reflected by the apple skin, which is the reason it appears red. And an object that absorbs all wavelengths and reflects none appears black.

The Scattering of Light

When sunlight encounters minute particles in the atmosphere, such as air or water molecules, or small particles of dust, it reflects off them in every direction. Sunlight is, in effect, bounced around like a pinball by these particles. This multi-directional **reflection** is called **scattering.**

Blue Skies

The scattering of sunlight by air molecules is what causes the sky to appear blue. However, it is a *selective scattering,* meaning that not all wavelengths of visible light are scattered equally. Air molecules scatter primarily violet, indigo, blue and green light, the colors at the short-wavelength end of the visible spectrum.

The small size of air molecules is responsible for the selective scattering of sunlight. The diameter of an air molecule is even smaller than the average wavelength of visible light. Air molecules, thus, are better able to scatter shorter wavelengths of visible light than longer wavelengths.

When you look at the sky, your eye is bombarded in all directions by violet, indigo, blue, and green light. However, the structure of the eye is such that the eye is much more sensitive to blue light than any of the other three colors. Thus, when violet, indigo, blue, and green light are present at once, what the eye perceives is blue.

If there were no air molecules or other particles in the air, in other words no scattering, the sky would appear black.

The Scattering of Light By Clouds

A cloud droplet has a far greater diameter than that of an individual air molecule. A cloud droplet, therefore, is capable of scattering all wavelengths of visible light to a fairly equal extent. In addition, a cloud droplet is a poor absorber of light. As we described above, an object that reflects the entire spectrum of visible light appears white.

Some of the sunlight that strikes a cloud will pass through the cloud, while the rest is reflected by the droplets it encounters. The sunlight that makes it through the cloud is reflected by droplets at the base, which is what makes the base appear white.

The amount of sunlight that penetrates a cloud depends on the thickness of the cloud. For instance, very little sunlight will reach the base of a cloud with a thickness of 3,300 feet (1,000 meters) or greater. That is the reason that the base of a tall cloud appears dark. The base of a cloud also darkens as its droplets become larger. The reason for this is that larger droplets of water are better at absorbing light than are smaller units. Therefore, the darkness of the base of a cloud is also an indication of the likelihood of rain.

Haze

Haze is the term used to describe a sky that has a uniform, milky white appearance. Haze is produced by high humidity in combination with a large number of particles in the air. Water vapor condenses around the suspended particles. These "haze particles," as they are called, scatter all wavelengths of visible light in all directions, just as cloud droplets do. The greater number of particles in the air, the whiter the sky appears.

The concentration of particles is often an indicator of air pollution, since the particles come mainly from emissions from smokestacks or automobile tailpipes (for more information on air pollution, see "Human Activity and the Future," page 508). Haze may also be created by naturally occurring particles in the air, such as pollen and dust.

Optical Effects

Haze occurs close to the surface. If you climb to the top of a tall mountain on a hazy day, you may see haze below and blue skies above.

CREPUSCULAR RAYS

Crepuscular rays are bright beams of light that radiate from the sun and cross the sky. They are most often visible at sunset or when the sun shines through a break in the clouds. The beams are made visible by the scattering of sunlight by dust, water droplets, or haze particles.

A similar effect is created when a bright light shines through a small opening, into a dusty room. Next time you're in a darkened movie theater, look toward the projection booth and notice the beam of projected light, which is being scattered by dust in the air.

Despite their appearance of fanning outward from the sun, crepuscular rays run parallel to one another. The fan shape is only an illusion, caused by perspective. This illusion is similar to that of a road, railroad track, or any other long straight path that appears to narrow to a single point in the distance.

COLORS AT SUNRISE AND SUNSET

We already explained why the sun appears white when it is high in the afternoon sky. So why is it that the sun appears red, yellow, or orange when it is on the horizon? The answer has to do with the angle at which

Crepuscular rays are sometimes referred to as "twilight rays," or alternating light and dark bands that appear to diverge in a fan-like array from the sun.

sunlight strikes a given location. In the middle of the day sunlight strikes the ground most directly, and at the beginning and end of the day sunlight strikes the ground at the steepest angle. The angle at which sunlight strikes the ground is indicative of the amount of atmosphere through which the sunlight must pass.

At sunrise and sunset, the sunlight must traverse the greatest distance of atmosphere. In fact, sunlight passes through about 12 more miles of atmosphere when the sun is just over the horizon, than it does when directly overhead. As the light of the setting or rising sun travels through all that atmosphere, its shorter wavelengths become scattered by the air molecules it encounters. The only wavelengths to make it all the way to Earth's surface are the longest wavelengths—red, orange, and yellow.

When the air is relatively clean, a setting or rising sun appears to be orange-yellow. An orange-red sun, however, indicates that the air contains a high concentration of particles. Particles that have diameters slightly larger than air particles scatter yellow wavelengths, leaving only light with the longest wavelengths—orange and red—to shine through. When the concentration of particles in the air is *very* high, such as after a volcanic eruption, only red light remains concentrated, and we see a red rising, or setting, sun.

There remains one more possibility as to how a rising or setting sun may appear: not at all. The disappearance of the sun before it reaches the horizon (in the absence of clouds) is the result of extreme air pollution. When the concentration of dust, smoke, and other pollutants in the air becomes great enough, even red wavelengths are scattered.

THE REFRACTION OF LIGHT

When light passes from one medium into a second medium, its speed changes. For example, when light travels from a less dense medium, say air, into a more dense medium, say water, the light slows down. And if the light enters the denser medium at any angle other than from straight above, it will bend. The bending of light, as it passes through two transparent media (plural form of *medium*) of different densities is called **refraction.** The degree to which light bends depends on both the densities of the two substances and the angle at which light enters the second substance.

When light is transmitted from a less dense substance to a more dense substance, it bends in the direction perpendicular to the boundary

Optical Effects

between the two substances. And when light is transmitted from a more dense substance to a less dense substance, it bends away from the perpendicular.

Positions of the Stars, Sun, and Moon

The light from the stars, sun, and moon travels from a less dense medium, space, to more dense medium, Earth's atmosphere. Thus, all starlight, sunlight, and moonlight, except that emitted when the sun, moon, or star is directly overhead, is refracted. As a result, our perception of the positions of the sun and stars is distorted.

When we look at a star in the sky, we are actually seeing the light from that star which has been bent upon entering our atmosphere. Our eyes cannot perceive that light has been bent, hence they cannot trace the path of light back to the actual position of the star. Therefore, the star appears to be higher in the sky than it actually is.

When a star is near the horizon, its light must pass through the greatest amount of atmosphere—and undergo the greatest amount of refraction before it reaches an observer on the surface—of any position in the sky. Thus, when a star is near the horizon, the star's image appears farthest from its true position.

Refraction also causes an observer to see the rising of the sun or moon about two minutes before it actually occurs, and the setting of the

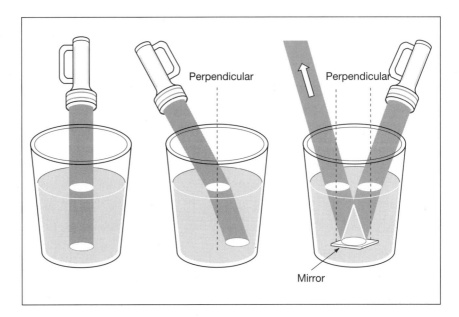

Figure 38: The refraction of light as it enters and exits water.

Optical Effects

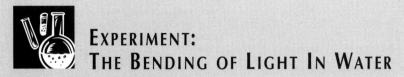

EXPERIMENT: THE BENDING OF LIGHT IN WATER

The best way to understand refraction is to try the following simple experiment: Fill a large glass beaker with water. Add a pinch of paprika or other ground spice. The spice causes a **scattering** effect, which makes the light beam easier to see. Then turn off the lights and shine a small flashlight into the water, from straight above. You will see the beam of light continue straight through to the bottom of the glass. Now tilt the flashlight so that the light enters the water at an angle. You will notice that the beam of light bends slightly toward the perpendicular.

sun or moon about two minutes after it actually occurs. This situation is similar to what happens with starlight on the horizon. Because the light of the sun or moon must shine through so much atmosphere, it is refracted in such a way that they appear higher in the sky than they really are.

The bending of light near the horizon is also responsible for the subtle "flattening" of a setting or rising sun or moon. Specifically, when the sun or moon is directly on the horizon, the light from the lower portion of the object is refracted to a greater degree than the light from the upper portion of the object. This causes the object to appear to be wider at the base than at the top.

GREEN FLASHES

A **green flash** is a very brief and difficult-to-see optical effect that accompanies a rising or setting sun. The green flash is a flash of green light that appears near the top edge of the sun. The green flash is due to both refraction and scattering of light in the atmosphere.

A green flash occurs because all wavelengths of light from a setting or rising sun are not refracted equally. Rather, the shorter wavelengths, purple and blue, are bent to the greatest degree, while the longer wavelengths, red and orange, are bent to the smallest degree.

Thus, we would expect that the color of the very tip of the sun as it first peeks over the horizon, or just sinks below the horizon, would be

Optical Effects

purple-blue. This would be true except that air molecules and tiny dust particles selectively scatter blue and purple wavelengths. The next longest wavelength, green, is what we see instead, unless there is a high concentration of particles in the air. In that case, green light will be scattered as well.

One reason that the green flash is so elusive is that in most cases, the green light is too faint to be seen by the human eye. The green flash is brightest, and noted most often, over the ocean where the air is relatively clean. It is also common at high latitudes, where the sun rises and sets more slowly. At the end of the long polar winter, the sunrise is so gradual that a green flash may persist for several minutes.

MIRAGES

Another product of the refraction of light in the atmosphere is the **mirage.** A mirage is an optical illusion such that an object appears in a position different from its true position. Alternatively, a nonexistent object, such as a body of water, appears. In some mirages, distant objects appear to be inverted or higher or lower than their true positions.

Mirages are caused by the refraction of light as it passes through layers of air with different densities. The differences in density are created by differences in temperature. The sharper the contrast in temperature between two air layers, the more pronounced the refraction of light.

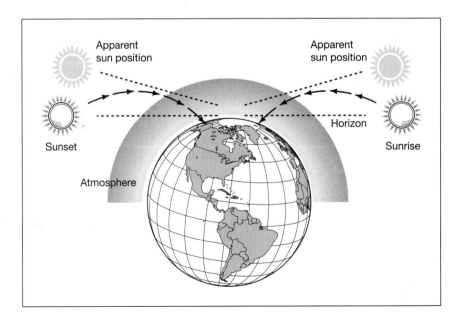

Figure 39: Refraction of sunlight as it travels through the atmosphere makes the sun appear to rise earlier, and set later, than it actually does.

Optical Effects

Perhaps the best-known mirage is the appearance of "water" commonly seen on a roadway or in the desert. This type of mirage forms on hot days, when the pavement or sand becomes very hot. Heat is transferred from the surface to the air immediately above it by **conduction** (see "What Is Weather?" on page 6). In contrast, the air just a few yards above the surface is cooler and denser.

A water mirage is formed by the displacement of blue light from sunlight. Specifically, the blue light of the sky near the horizon is not reflected directly from the sky to the ground. Rather, as it crosses the cool air/warm air boundary, it is refracted upward, toward our eyes. Our eyes thus perceive the blue sky light as coming from just above the surface, in the distance. As we move toward the location where we saw the mirage, the mirage disappears. The mirage re-appears ahead, moving with us and always remaining in the distance.

Water mirages are even more convincing when they appear to "shimmer," as does water when sunlight strikes it. The shimmering of a water mirage is caused by small shifts in the degree to which light is refracted. The reason why the angle of refraction changes is that near the surface hot air is constantly rising and cool air is constantly sinking. This variation causes a continual change in density of air layers which, in turn, causes the continual shift in the amount by which light is refracted.

In this mirage seen from Savoonga, Alaska, Siberian mountains are refracted to appear much closer than they really are.

Optical Effects

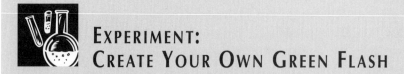

EXPERIMENT: CREATE YOUR OWN GREEN FLASH

To learn how a green flash is created, perform this simple experiment. All it requires is a prism, a lamp (minus the shade), and a piece of dark paper. Hold the prism so it is between yourself and the light. You will see the entire spectrum of colors in the prism, with red on the bottom and blue on the top.

Then take the piece of dark paper and move it up, gradually, between the prism and the light. First red will be blocked out, then orange, yellow, green, blue, and finally violet. As we described above, in the atmosphere, blue light is reflected in all directions by air molecules. Thus, the shortest-wavelength color we would see refracted during a sunset would be green.

INFERIOR MIRAGES. On a hot day, when you look at an object in the distance, you may see an upside-down version of the object directly beneath it. Yet, when you get right up to the object, the upside-down image disappears. The upside-down image is an optical illusion called an **inferior mirage.** Inferior mirages are similar to the water mirages just described. They form when the surface air is hotter and less dense than air at higher elevations.

Light is reflected outward from a distant object—a tree, for example—in all directions. Some of the light from the treetop travels a straight horizontal path, never dipping into the warmer surface air. When that light reaches your eye, you get an accurate image of the tree. The light from the lower portions of the tree is not refracted because it remains within a medium of a single density, between the tree and your eye.

However, some of the light from the treetop travels at a gentle slope downward and eventually crosses the boundary between air layers. When it does so, it bends upward. When the refracted light reaches your eye, your eye follows the path of the light, at its refracted angle, back over the distance to the tree. Thus, the image appears lower to the ground than it actually is.

The light from the top of the tree is bent upward to the greatest degree, making it appear to come from the lowest position. For this reason, the lowered image of the tree is also upside down.

SUPERIOR MIRAGES. Superior mirages are created under conditions that are opposite of those that create inferior mirages. Namely, they form in cold weather when the surface air is colder, and thus denser, than the air above. In a superior mirage, a distant object appears to be taller and closer to the observer than it actually is. Sometimes it appears upside down. Superior mirages are most common in polar regions, where the air over a snow-covered surface is colder than the air several feet above.

Let's take the example of a mountain in the distance. The light from the mountaintop is reflected in all directions. Some of that light follows a gently sloping path downward. When that light enters the layer of colder air, it is bent toward the perpendicular, in this case, into a steeper downward path. When this refracted light reaches your eye, your eye follows the path of the light, at its refracted angle, back over the distance to the mountain. The image thus appears higher above the ground than it actually is.

A special type of superior mirage is called a **Fata Morgana.** A Fata Morgana takes the form of spectacular castles, buildings, or cliffs, rising above cold land or water, particularly in polar regions. This type of mirage is produced by light that is refracted as it passes through air layers of various temperatures. A Fata Morgana requires that the air temperature over a cold surface increase with height. Specifically, the temperature rises slowly throughout the surface layer of air, then several feet above

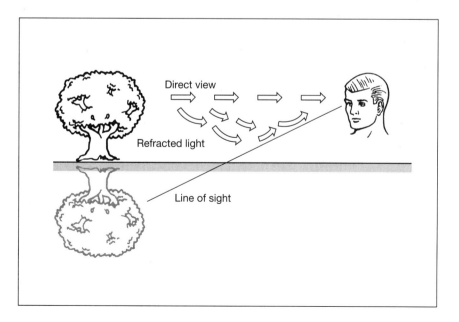

Figure 40: Inferior mirage of a tree over a hot surface.

Optical Effects

the surface the air temperature rises more quickly. In the next layer of air, the temperature rises slowly again.

HALOS

A **halo** is a thin ring of light that appears around the sun or the moon. Halos are caused by the refraction of light by ice crystals. These ice crystals are either free-falling or within upper-level clouds called **cirriform** clouds (see "Clouds," page 75). Cirriform clouds are the only type that are both high enough to contain ice crystals, yet thin enough to allow the image of the sun to shine through.

There are two main types of halo: the 22° halo and the 46° halo. The 22° halo is smaller, more tightly encircles the sun, and is more common, than the 46° halo. Several other sizes of halo appear very infrequently. We will not go into detail on halos other than 22° and 46° in this chapter.

The size of a halo (22° or 46°) refers to the angle by which light is refracted through ice crystals and, consequently, the radius of the halo. For instance, if light is refracted by ice crystals at an angle of 22°, it will form a circle of light with a radius of 22°. To better understand this, draw a picture of a person on the ground and a moon with a halo, above. Draw two lines: one between the person and the moon, and the other between the person and a point on the far left or far right side of the halo. The angle formed by the two lines, which in reality would either be 22° or 46°, indicates whether the halo has a radius of 22° or 46°.

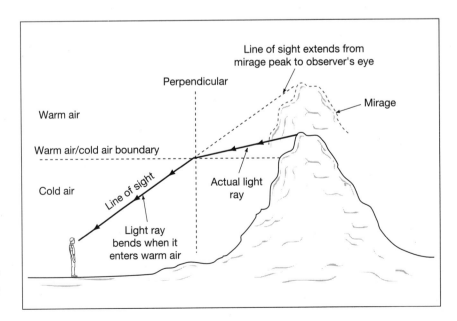

Figure 41: Superior mirage of a distant mountain.

Both 22° and 46° types of halo are formed when light strikes small, pencil-shaped, hexagonal ice crystals that are around 20 micrometers (about 8 ten-thousandths, or .0008, of an inch) in diameter. The ice crystals that form a 46° halo may be as small as 15 micrometers or as large as 25 micrometers, while the ice crystals that form a 22° halo are more uniformly 20 micrometers.

A 22° halo is the result of refraction by randomly oriented ice crystals. The light enters one of the six sides and exits through another of the six sides. In the process, the light is bent by an angle of 22°.

The ice crystals that produce a 46° halo are oriented in such a way that the sun strikes one of the six sides and exits through one of the two ends. This arrangement causes the sunlight to be refracted at an angle of 46°.

Haloes may form at the leading edge of an advancing **frontal system.** Thus, they are often looked upon as a sign of rain. A halo is certainly not a foolproof forecasting tool, however, because the front may change direction or gently pass through without producing rain.

SUNDOGS

Sundogs are also called mock suns or perihelia, Greek for "beside the sun." They consist of one or two patches of light that appear on either or both sides of the sun. Sundogs make it appear that there are two or

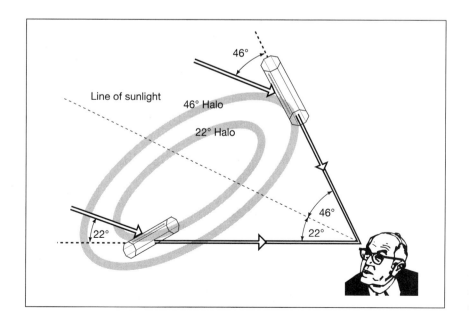

Figure 42: 22° and 46° halos are formed by the refraction of sunlight (or moonlight) by ice crystals at different orientations.

Optical Effects

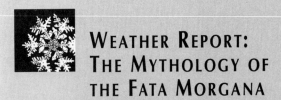

WEATHER REPORT: THE MYTHOLOGY OF THE FATA MORGANA

The name "Fata Morgana" is Italian for "Fairy Morgan." According to mythology, Fairy Morgan was King Arthur's half-sister. She lived in an underwater crystal palace and was capable of creating magical castles out of thin air. In the fifteenth century, Italian poets from the town of Reggio viewed a fantastic, castle-like mirage near the Strait of Messina (the waterway between Italy and Sicily). Unable to explain what they saw, they called it a "Fata Morgana," and the name stuck.

three suns in the sky. When two sundogs occur, one may be brighter than the other, or higher than the other. They may appear white or colored. Sundogs often appear along the circumference of a 22° halo.

Occasionally these patches of light are seen around a very bright, full moon. In that case, they are called **moon dogs.**

As illustrated in Figure 43, sundogs are produced by the refraction of sunlight that shines through plate-like ice crystals with diameters of at least 30 micrometers (.0012 inch). Since these ice crystals must be present near the ground, sundogs occur only in cold regions.

The plate-like ice crystals fall slowly through the air and are randomly oriented. Only those ice crystals that are positioned horizontally, with large, flat ends parallel to the ground, will refract sunlight at an angle of 22°. It takes millions of falling ice crystals, all oriented so that they refract sunlight at 22°, to produce sundogs.

Where these falling ice crystals are relatively large and plentiful, the sundogs will be colorful. This color is produced by the *selective refraction* of light, also called **dispersion.** In the process of dispersion, each ice crystal acts like a tiny prism, separating sunlight into the spectrum of colors.

The amount by which each color is refracted by an ice crystal varies slightly. Red light has the longest wavelength and is slowed the least as it passes through the ice crystal. Hence, red is bent the least. On the other extreme, violet light has the shortest wavelength and is slowed the most as it passes through the ice crystal. Hence, violet is bent the most.

The result is that red light appears on the edge of the sundog closest to the sun and blue appears on the edge farthest from the sun. The reason why blue, and not purple, appears is that the human eye is better able to perceive blue than purple.

Occasionally a halo will also be colorful, rather than its characteristic white. This dispersion of sunlight into bands of color, by the process just described, occurs when the ice crystals are relatively large and of uniform size and shape.

Rainbows

A **rainbow** is an arc of light, separated into its constituent colors, that stretches across the sky. Rainbows are products of both **reflection** and **refraction** of sunlight by raindrops. A rainbow is, in effect, sunlight that has undergone **dispersion** and is reflected back to your eye. To observe a rainbow, the sun must be at your back and the falling rain must be in front of you.

A rainbow is formed by a rather complex process, as illustrated in Figure 44. As sunlight enters a raindrop, it is dispersed into its constituent colors, meaning that each color of the spectrum is refracted to a

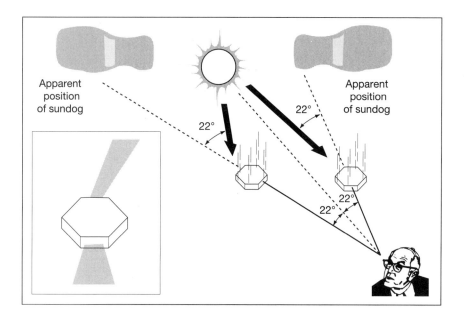

Figure 43: Sundogs are produced by the refraction of sunlight through falling, plate-like crystals.

Optical Effects

different degree. Most of this dispersed sunlight passes right through the raindrop. However, when sunlight strikes the back of the raindrop at a certain angle, called the **critical angle,** the sunlight is reflected back to the front of the drop. To achieve this critical angle, the sun can be no higher than 42 degrees above the horizon.

As a result of dispersion, once the sunlight enters the raindrop, each color strikes the back of the raindrop at a slightly different angle. Thus, each color reflects off the back of the raindrop and emerges from the front of the raindrop at a slightly different angle.

Only one color exits from each raindrop at the exact angle necessary to reach your eye. This means that you see only one color at a time reflecting from each raindrop. For this reason, it takes millions of raindrops to create a rainbow.

Due to its angle of refraction, red light is reflected to your eye from the highest raindrops. Therefore, red is the color at the top edge of the rainbow. And violet light, which is reflected from the lowest raindrops, forms the bottom edge of the rainbow. The rest of the spectrum—orange, yellow, green, blue, and indigo—fills in the middle portion of the rainbow.

Each time you move, the rainbow you observe is being reflected from a whole different set of raindrops. Each raindrop produces only one ray of light at the appropriate angle to intercept your eye. By the same token, no two people can observe exactly the same rainbow!

Sundogs formed by high level ice crystal clouds.

Sometimes two rainbows appear in the sky at once. The brighter rainbow, formed by the process just described, is the *primary rainbow*. The fainter rainbow is called the *secondary rainbow*.

A secondary rainbow is formed when sunlight strikes the raindrops at such an angle that the light is reflected twice within each drop. This double reflection causes violet light to be reflected to the eye from higher raindrops and red light, from lower raindrops. Therefore, in the secondary rainbow, the order of colors is reversed, with violet on top and red on the bottom. Some light is lost in the double-reflection process, which is the reason the secondary rainbow is dimmer than the primary one.

THE DIFFRACTION OF LIGHT

Diffraction is the slight bending of sunlight or moonlight around water droplets or other tiny particles that it encounters. The diffraction of

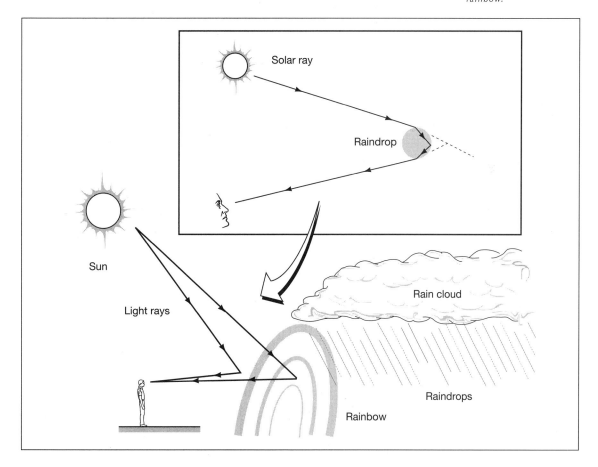

Figure 44: The formation and location of a rainbow.

Optical Effects

sunlight or moonlight produces patches of white and colored light in the sky.

CORONAS

A **corona** (Latin for "crown") is a circle of light centered on the moon or sun that is usually bounded by a colorful ring or set of rings. Coronas are rarely observed around the sun because of the sun's brightness. Therefore, we will limit our description of coronas in this section to lunar coronas.

A corona is the product of the diffraction of moonlight around tiny, spherical cloud droplets. A corona can form only on nights when the moon is visible through a thin layer of clouds.

The smaller the cloud droplets, the greater the angle of diffraction, and the larger the corona. The largest coronas are produced by a newly formed cloud of uniform thickness. Coronas are much smaller than **halo**s (see page 328), however, because the angle of diffraction that produces a corona is only a few degrees.

Sometimes alternating light and dark bands are visible in the middle portion of a coronas. Light and dark bands are formed when light waves, which have bent around a water droplet, come back together in particular ways.

Light waves are similar to water waves, in that they have crests and troughs. When the crest of one wave meets the crest of another wave, the two are added together and become one large wave. This phenomena is called *constructive interference*. In light waves, constructive interference produces a bright band.

A double rainbow.

The opposite of constructive interference is *destructive interference*. Destructive interference occurs when the crest of one wave meets the trough of another and they cancel each other out. When destructive interference occurs in water waves, a calm spot is produced. When it occurs in light waves, a dark band is produced.

Now we come to the colored rings on the edges of a corona. These rings are produced by diffraction of moonlight around cloud droplets of uniform size. If the droplets are different sizes, the color will appear in an irregular, and not a circular, pattern.

In a process similar to that of **dispersion** (see page 331), diffraction causes the differential bending of light, according to wavelength. When white light is bent, its longest-wavelength component, red, bends the least and its shortest-wavelength component, violet, bends the most. In this way the light is separated into its constituent colors.

Red appears on the side of the ring farthest from the moon and violet appears on the side of the ring closest to the moon. The corona may have several rings, which become fainter with distance from the moon.

Iridescence

Iridescence is the term used to describe irregular patches of colored light on clouds. This effect is most often seen within 20 degrees of the sun or moon. Iridescence often appears as pastel shades of blue, pink, or green. The brightness of the colors is proportional to the number of droplets within the cloud and the uniformity of size of those droplets.

Iridescence forms in the same way as a corona and is, essentially, an irregular corona. One difference between the two phenomena has to do with the size of the cloud droplets. Iridescence is formed when sunlight is diffracted by cloud droplets of different sizes, while coronas require cloud droplets of a uniform size.

Sometimes iridescence appears as an arc, or a portion of a corona. This is the case when a cloud partially obscures the sun or moon, but does

A rare solar corona, caused by ice crystals from thin cirrostratus clouds ringing the sun.

Optical Effects

not cover the entire region in which a corona would form. Iridescence may also form on a cloud that is near, but not covering, the sun or moon.

GLORIES

A **glory** is a set of colored rings that appears on the top surface of a cloud, directly beneath an observer. Although it is possible to view a glory by climbing a mountain until you're above the clouds, it's much easier to view one from an airplane window. Because they are most often viewed from airplanes, glories are generally thought of as the rings of color that surround the shadow of an airplane.

A glory is formed by a rather complex process, similar to the formation of a **rainbow** (see page 331). The main difference between the two phenomena, however, is that rainbows are formed by the interaction of sunlight with raindrops while glories are formed by the interaction of sunlight with tiny cloud droplets. The cloud droplets are less than 50 micrometers (about .002 of an inch) in diameter. In contrast, raindrops are .04 to .24 inches (.1 to .6 centimeters) in diameter, on average.

In a glory, sunlight undergoes **refraction, reflection,** and **diffraction** within cloud droplets before being returned to your eye. First, sunlight that strikes the surface of a cloud droplet is refracted within the droplet. This refracted light then reflects off the back of the droplet. Some of this reflected light skims the opposite surface of the droplet and

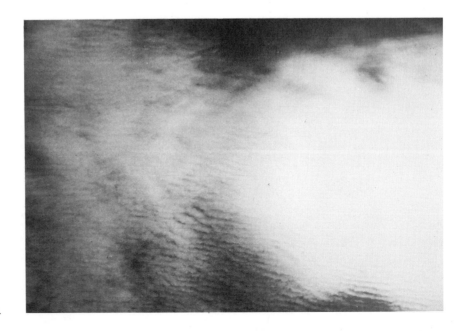

Iridescent clouds.

bends slightly, or is diffracted, around the droplet. The light then exits the droplet on a path that is parallel to its entry path.

The process of diffraction is also what separates the light of a glory into its constituent colors. As the light is diffracted by cloud droplets, red is bent the least and violet, the most. Hence, as with a rainbow, each droplet reflects light of only one color. The innermost ring of the glory appears purple and the outermost ring appears red, with the rest of the spectrum lying in between.

A glory is always positioned directly beneath the observer. And, like a rainbow, a glory moves with the observer. Thus, if you were on one airplane and another airplane was flying beside you, you would be able see the glory around your own plane's shadow, but not the glory around the other plane's shadow.

AURORAS

Here we will take a look at the only type of optical effects mentioned thus far that are not produced by sunlight or moonlight: **auroras**. Auroras are bright, colorful displays of light in the night sky. They come in two forms—aurora borealis and aurora australis—better known as the northern and southern lights. Auroras are most prominent near the North and South poles, but can be seen occasionally in the **middle latitudes.**

The aurora borealis, or northern lights, put on a beautiful show.

Optical Effects

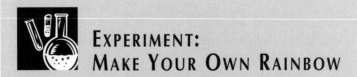

EXPERIMENT: MAKE YOUR OWN RAINBOW

You can create your own rainbow using a clear glass bowl of water, a flashlight, and a small, flat mirror. The water acts as a refractor and the mirror acts as a reflector.

Simply place the mirror in the bowl of water, so that it rests against the side of the bowl at about a 45° angle. Then shine a flashlight straight down at the mirror. A rainbow will appear on the wall opposite the mirror.

A display of northern or southern lights can be as fascinating as fireworks. It varies in color from whitish-green to deep red and takes on shapes such as streamers, arcs, curtains, and shells.

Auroras are produced when charged particles from the sun enter Earth's atmosphere. As this stream of particles approaches Earth, it is trapped for a time in the outermost parts of Earth's magnetic field. Eventually the particles are drawn down toward the north and south magnetic poles. Along the way, they ionize (create an electric charge within) oxygen and nitrogen gas in the atmosphere. This causes the atmosphere to glow.

Sources

Books

Ahrens, C. Donald. *Meteorology Today: An Introduction to Weather, Climate, and the Environment.* 5th ed. St. Paul, MN: West Publishing Company, 1994.

Allaby, Michael. *How the Weather Works: 100 Ways Parents and Kids Can Share the Secrets of the Atmosphere.* Pleasantville, NY: The Reader's Digest Association, Inc., 1995.

Allaby, Michael, ed. *Illustrated Dictionary of Science.* Rev. ed. New York: Facts on File, Inc., 1995.

Anthes, Richard A. *Meteorology.* 6th ed. New York: Macmillan Publishing Company, 1992.

Bair, Frank E., ed. *The Weather Almanac.* 6th ed. Detroit: Gale Research, 1992.

Bair, Frank E., ed. *Weather of U.S. Cities.* 4th ed. Detroit: Gale Research, 1992.

Burroughs, William J. and Bob Crowder, et. al. *Nature Company Guides: Weather.* New York: Time Life Books, 1996.

Burroughs, William James. *Watching the World's Weather.* Cambridge, England: Cambridge University Press, 1991.

Carnegie Library of Pittsburgh, Science and Technology Department. *The Handy Science Answer Book.* Detroit: Visible Ink Press, 1994.

Christian, Spencer. *Spencer Christian's Weather Book.* New York: Prentice Hall, 1993.

Climates of the States. 3rd ed. Detroit: Gale Research, 1985.

Cosgrove, Brian. *Eyewitness Books: Weather.* New York: Alfred A. Knopf, 1991.

Davies, J. K. *Space Exploration.* Edinburgh, Scotland: W & R Chambers Ltd., 1992.

Day, John A. and Vincent Schaefer. *Peterson First Guide to Clouds and Weather.* New York: Houghton Mifflin Company, 1991.

DeMillo, Rob. *How Weather Works.* Emeryville, CA: Ziff-Davis Press, 1994.

Dolan, Edward F. *The Old Farmer's Almanac: Book of Weather Lore.* Dublin, NH: Yankee Publishing Inc., 1988.

Dunlop, S. and F. Wilson. *The Larousse Guide to Weather Forecasting.* New York: Larousse and Co. Inc., 1982.

Engelbert, Phillis. *Astronomy and Space: From the Big Bang to the Big Crunch.* Detroit: U•X•L, 1997.

Sources

Fields, Alan. *Partly Sunny: The Weather Junkie's Guide to Outsmarting the Weather.* Boulder, CO: Windsor Peak Press, 1995.

Fishman, Jack and Robert Kalish. *The Weather Revolution: Innovations and Imminent Breakthroughs in Accurate Forecasting.* New York: Plenum Press, 1994.

Gaskell, T. F. and Martin Morris. *World Climate: The Weather, the Environment and Man.* New York: Thames and Hudson, 1979.

Graedel, Thomas E. and Paul J. Crutzen. *Atmosphere, Climate, and Change.* New York: Scientific American Library. 1995.

Greenler, Robert. *Rainbows, Halos, and Glories.* Cambridge, England: Cambridge University Press, 1980.

Hardy, Ralph. *Teach Yourself Weather.* Lincolnwood, IL: NTC Publishing Group, 1996.

Hardy, Ralph, et al. *The Weather Book.* Boston: Little, Brown and Company, 1982.

Hatherton, Trevor, ed. *Antarctica: The Ross Sea Region.* Wellington, New Zealand: DSIR Publishing, 1990.

Henson, Robert. *Television Weathercasting: A History.* Jefferson, NC: McFarland & Company, Inc., 1990.

Holford, Ingrid. *Weather Facts & Feats.* 2nd ed. Middlesex, England: Guinness Superlatives Limited, 1982.

Kahl, Jonathan D. W. *Weather Watch: Forecasting the Weather.* Minneapolis: Lerner Publications Company, 1996.

Keen, Richard A. *Michigan Weather.* Helena, MT: American & World Geographic Publishing, 1993.

Lamb, H. H. *Climate, History and the Modern World.* 2nd ed. London, England: Routledge, 1995.

Lambert, David and Ralph Hardy. *Weather and Its Work.* London, England: Orbis Publishing Limited, 1984.

Lampton, Christopher. *Meteorology: An Introduction.* New York: Franklin Watts, 1981.

Leggett, Jeremy, ed. *Global Warming: The Greenpeace Report.* Oxford, England: Oxford University Press, 1990.

Lockhardt, Gary. *The Weather Companion: An Album of Meteorological History, Science, Legend, and Folklore.* New York: John Wiley & Sons, Inc., 1988.

Ludlam, F. H. and R. S. Scorer. *Cloud Study: A Pictorial Guide.* London, England: John Murray, 1957.

Ludlum, David M. *The Weather Factor.* Boston: Houghton Mifflin Company, 1984.

Lutgens, Frederick K. and Edward J. Tarbuck. *The Atmosphere: An Introduction to Meteorology.* 5th ed. Englewood Cliffs, NJ: Prentice Hall, 1992.

Lydolph, Paul E. *The Climate of the Earth.* Lanham, MD: Rowman & Littlefield Publishers, Inc., 1985.

Lynch, David K. and William Livingston. *Color and Light in Nature.* Cambridge, England: Cambridge University Press, 1995.

Lyons, Walter A. *The Handy Weather Answer Book.* Detroit: Visible Ink Press, 1997.

Mason, Helen. "Tom Kudloo: Aerologist." *Great Careers for People Who Like Being Outdoors.* Detroit: U•X•L, 1993.

McPeak, William J. "Edward N. Lorenz." *Notable Twentieth-Century Scientists.* Vol. 3. Ed. Emily J. McMurray. New York: Gale Research, 1995.

McPeak, William J. "Jacob Bjerknes." *Notable Twentieth-Century Scientists.* Vol. 1. Ed. Emily J. McMurray. New York: Gale Research, 1995.

Sources

Meinel, Aden and Marjorie Meinel. *Sunsets, Twilights, and Evening Skies.* Cambridge, England: Cambridge University Press, 1983.

Mogil, H. Michael and Barbara G. Levine. *The Amateur Meteorologist: Explorations and Investigations.* New York: Franklin Watts, 1993.

Moran, Joseph M. and Lewis W. Morgan. *Essentials of Weather.* Englewood Cliffs, NJ: Prentice Hall, 1995.

Moran, Joseph M. and Lewis W. Morgan. *Meteorology: The Atmosphere and the Science of Weather.* Edina, MN: Burgess Publishing, 1986.

Naseri, Muthena and Douglas Smith. "Solar Energy." *Environmental Encyclopedia.* Cunningham, William P., et al, eds. Detroit: Gale Research, 1994.

National Research Council, Committee on Atmospheric Sciences. *Weather and Climate Modification.* Detroit: Grand River Books, 1973.

Newton, David E. *Global Warming: A Reference Handbook.* Santa Barbara, CA: ABC-CLIO, Inc., 1993.

Newton, David E. *The Ozone Dilemma: A Reference Handbook.* Santa Barbara, CA: ABC-CLIO, Inc., 1995.

Newton, David E. "Wind Energy." *Environmental Encyclopedia.* Cunningham, William P., et al, eds. Detroit: Gale Research, 1994.

Oleck, Joan. "Tetsuya Theodore Fujita." *Notable Twentieth-Century Scientists.* Vol. 2. Ed. Emily J. McMurray. New York: Gale Research, 1995.

Pine, Devera. "Carl-Gustaf Rossby." *Notable Twentieth-Century Scientists.* Vol. 3. Ed. Emily J. McMurray. New York: Gale Research, 1995.

Posey, Carl A. *The Living Earth Book of Wind & Weather.* Pleasantville, NY: The Reader's Digest Association, Inc., 1994.

Robinson, Andrew. *Earth Shock: Hurricanes, Volcanoes, Earthquakes, Tornadoes and Other Forces of Nature.* New York: Thames and Hudson, 1993.

Roth, Charles E. *The Sky Observer's Guidebook.* New York: Prentice Hall Press, 1986.

Rubin, Louis D. and Jim Duncan. *The Weather Wizard's Cloud Book.* Chapel Hill, NC: Algonquin Books of Chapel Hill, 1989.

Ryan, Martha. *Weather.* New York: Franklin Watts, 1976.

Schaefer, Vincent J. and John A. Day. *A Field Guide to the Atmosphere.* Boston: Houghton Mifflin Company, 1981.

Schneider, Stephen H., ed. *Encyclopedia of Climate and Weather.* New York: Oxford University Press, 1996.

Schwarz, Joel. "Lewis Fry Richardson." *Notable Twentieth-Century Scientists.* Vol. 3. Ed. Emily J. McMurray. New York: Gale Research, 1995.

Scorer, Richard. *Clouds of the World.* Melbourne, Australia: Lothian Publishing Co., 1972.

Scorer, Richard and Arjen Verkaik. *Spacious Skies.* Newton Abbot, England: David & Charles Publishers, 1989.

Sybil, P. Parker, ed. *McGraw-Hill Dictionary of Earth Science.* 5th ed. New York: McGraw-Hill, 1997.

Tannenbaum, Beulah and Harold E. Tannenbaum. *Making and Using Your Own Weather Station.* New York: Franklin Watts, 1989.

Travers, Bridget, ed. *The Gale Encyclopedia of Science.* 6 Volumes. New York: Gale Research, 1996.

Travers, Bridget, ed. *World of Invention.* Detroit: Gale Research, 1994.

Travers, Bridget, ed. *World of Scientific Discovery.* Detroit: Gale Research, 1994.

Sources

Trewartha, Glenn T. and Lyle H. Horn. *An Introduction to Climate*. 5th ed. New York: McGraw-Hill Book Co., 1980.

Vickery, Donald M. and James F. Fries. *Take Care of Yourself*. 6th ed. Reading, MA: Addison-Wesley Publishing Company, 1996.

Wagner, Ronald L. and Bill Adler, Jr. *The Weather Sourcebook: Your One-Stop Resource for Everything You Need to Feed Your Weather Habit*. Old Saybrook, CT: The Globe Pequot Press, 1994.

Watt, Fiona and Francis Wilson. *Weather and Climate*. London: Usborne Publishing Ltd., 1992.

Williams, Jack. *The Weather Book: An Easy-to-Understand Guide to the USA's Weather*. New York: USA Today & Vintage Books, 1992.

Witze, Alexandra. "Alfred Wegener." *Notable Twentieth-Century Scientists*. Vol. 4. Ed. Emily J. McMurray. New York: Gale Research, 1995.

World Meteorological Organization. *International Cloud Atlas*. Geneva, Switzerland: World Meteorological Organization, 1987.

Articles

Addison, Doug. "Filmmaker to Storm Chasers." *Weatherwise*. June/July 1996: 29–32.

Addison, Doug. "Superstorm Success." *Weatherwise*. June/July 1995: 18–24.

Addison, Doug. "Weathercasting: Forecasts on the Fly." *Weatherwise*. August/September 1995: 8–9.

Bentley, Mace. "A Midsummer's Nightmare." *Weatherwise*. August/September 1996: 13–19.

Black, Harvey. "Heat: Air Mass Murderer." *Weatherwise*. August/September 1996: 11–12.

Black, Harvey. "Hurricanes: Satellite Enhancements." *Weatherwise*. February/March 1996: 10–11.

Brotak, Edward. "Reviews and Resources." *Weatherwise*. October/November 1996: 40–41.

Brotak, Edward, Stanley Gedzelman, and Dean Lewis. "Reviews and Resources." *Weatherwise*. December 1996/January 1997: 47.

Browne, Malcolm W. "Are Lightning Balls Spheres of Plasma?" *The New York Times*. 10 September 1996: C1, C9.

Cerveny, Randy. "From Corn Flakes to Computers: Making Weather in the Movies" *Weatherwise*. December 1996/January 1997: 35–40.

"Carolinas Clean Up the Mess Bertha Left." *The Washington Post*. 14 July 1996: A4.

Coco, Mark J. "Stalking the Green Flash!" *Weatherwise*. December 1996/January 1997: 31–34.

Corfidi, Stephen. "The Colors of Twilight." *Weatherwise*. June/July 1996: 14–19.

Darack, Ed. "Majestic Mantle: Mountain Weather Can Be a Climber's Reward." *Weatherwise*. December 1995/January 1996: 24–28.

De Wire, Elinore. "England's Great Storm." *Weatherwise*. October/November 1996: 34–38.

De Wire, Elinore. "When the Heavens Dance: The Awesome Aurora Inspire Humanity To Create Magical Metaphor." *Weatherwise*. December 1995/January 1996: 18–20.

Dickinson, Robert. "The Climate System." *Reports to the Nation on Our Changing Planet*. Washington: NOAA, Winter 1991.

Eames, Stanley B. "Project Atmosphere: Improving the Teaching of Meteorology."

Sources

Weatherwise. August/September 1992: 20–23.

Fields, Alan. "A Gift Fit for a Weather Buff: Birthday Shopping Made Easy." *Weatherwise.* December 1995/January 1996: 22–23.

Gedzelman, Stanley. "Automating the Atmosphere." *Weatherwise.* June/July 1995: 46–51.

Gedzelman, Stanley. "Beyond Bergen." *Weatherwise.* June/July 1995: 37.

Gedzelman, Stanley. "Halo Heaven: Close Encounters With Colorful Rings." *Weatherwise.* August/September 1995: 34–40.

Gedzelman, Stanley. "Mysteries in the Clouds." *Weatherwise.* June/July 1995: 55–57.

Gedzelman, Stanley. "Our Global Perspective." *Weatherwise.* June/July 1995: 63–67.

Gedzelman, Stanley. "Using Your Computer: The Power of Imitation." *Weatherwise.* April/May 1993: 36+.

Gedzelman, Stanley. "Using Your Computer: Weaving Rainbows." *Weatherwise.* August/September 1996: 42–45.

Gedzelman, Stanley and Patrick Hughes. "The New Meteorology." *Weatherwise.* June/July 1995: 26–36.

Geer, Ira W. *Increasing Weather Awareness with NOAA Weather Radio: A Guide for Elementary Schools.* Washington: GPO, 1983.

Graf, Dan. "California Crazy: West Coast Winters Are Often Far From Laid-Back When El Niño Flares Up." *Weatherwise.* December 1995/January 1996: 29–32.

Graf, Daniel, William Gartner and Paul Kocin. "Snow." *Weatherwise.* February/March 1996: 48–52.

Grenci, Lee. "When Storms Die." *Weatherwise.* June/July 1996: 48–49.

Henson, Robert. "Smells Like Rain." *Weatherwise.* April/May 1996: 29–32.

Hickcox, David H. "Temperature Extremes." *Weatherwise.* February/March 1996: 54–58.

Hill, Carolinda. "Mayday!" *Weatherwise.* June/July 1996: 25–28.

Horstmeyer, Steve. "Tilting at Wind Chills: Is Winter's Popular Index Blown Out of Proportion?" *Weatherwise.* October/November 1995: 24–28.

Hughes, Patrick. "Dust Bowl Days." *Weatherwise.* June/July 1995: 32–33.

Hughes, Patrick. "The Meteorologist in Your Life." *Weatherwise.* June/July 1995: 68–71.

Hughes, Patrick. "Probing the Sky." *Weatherwise.* June/July 1995: 52–54.

Hughes, Patrick. "Realizing the Digital Dream." *Weatherwise.* June/July 1995: 44–45.

Hughes, Patrick. "The View from Space." *Weatherwise.* June/July 1995: 60–62.

Hughes, Patrick. "Winning the War." *Weatherwise.* June/July 1995: 38–41.

Hughes, Patrick and Douglas Le Comte. "Tragedy in Chicago." *Weatherwise.* February/March 1996: 18–20.

Hughes, Patrick and Richard Wood. "Hail: The White Plague." *Weatherwise.* April/May 1993: 16–21.

Iocavelli, Debi. "Hurricanes: Eye Spy." *Weatherwise.* August/September 1996: 10–11.

Kristof, Nicholas D. "In Pacific, Growing Fear of Paradise Engulfed." *The New York Times.* 2 March 1997: 1.

Le Comte, Douglas. "Going to Extremes: 1995 Was Wild and Woolly for the U.S." *Weatherwise.* February/March 1996: 14+.

Sources

Martner, Brooks. "An Intimiate Look at Clouds." *Weatherwise*. June/July 1996: 20–22.

Marshall, Steve. "Hortense Packs a Punch, but USA Might be Spared." *USA Today*. 12 September 1996: A3.

Marshall, Tim. "A Passion for Prediction: There's More To Chasing Than Intercepting a Tornado." *Weatherwise*. April/May 1993: 22–26.

Mayfield, Max and Miles Lawrence. "Atlantic Hurricanes." *Weatherwise*. February/March 1996: 34–41.

McDonald, Kim A. "Preserving a Priceless Library of Ice." *The Chronicle of Higher Education*. 2 August 1996: A7+.

McDonald, Kim A. "Tornado-Chasing Scientists Use New Techniques to Probe the Origins of the Deadly Storms." *The Chronicle of Higher Education*. 12 July 1996: A9+.

McDonald, Kim A. "Unearthing Earth's Ancient Atmosphere Beneath Two Miles of Greenland Ice." *The Chronicle of Higher Education*. 2 August 1996: A6+.

Meredith, Robyn. "In Ohio River Valley, the Water's Edge Is Now It's Middle." *The New York Times*. 6 March 1997: A14.

Mervis, Jeffrey. "Agencies Scramble to Measure Public Impact of Research." *Science*. 5 July 1996: 27–28.

Mogil, H. Michael and Barbara G. Levine. "Gallery of Weather Pages." *Weatherwise*. August/September 1995: 15–16.

National Oceanic and Atmospheric Administration. *Are You Ready for a Winter Storm?* Washington: NOAA, 1991.

National Oceanic and Atmospheric Administration. *Heat Wave*. Washington: NOAA, 1994.

Nielsen, Clifford H. "Hurd Willett: Forecaster Extraordinaire." *Weatherwise*. August/September 1993: 38–44.

Pettengill, Steve. "Sailing by Satellite." *Weatherwise*. October/November 1995: 17–22.

Rice, Doyle. "Olympics: Gold Medal Forecasts." *Weatherwise*. June/July 1996: 10–12.

Richards, Steven J. "Hail To the Bronx: Using Weather To Turn City Kids On To Science." *Weatherwise*. August/September 1992: 24–28.

Rosenfeld, Jeff. "Cars vs. the Weather: A Century of Progress." *Weatherwise*. October/November 1996: 14–23.

Rosenfeld, Jeff. "Excitement in the Air." *Weatherwise*. June/July 1995: 71–72.

Rosenfeld, Jeff. "The Forgotten Hurricane." *Weatherwise*. August/September 1993: 13–18.

Rosenfeld, Jeff. "The Jumbo Outbreak." *Weatherwise*. June/July 1995: 58–59.

Rosenfeld, Jeff. "Spin Doctor: Talking Tornadoes with Howard Bluestein." *Weatherwise*. April/May 1996: 19–25.

Ryan, Bob. "A Window on Science: Watching the Weather Brings Out the Scientist In Everyone." *Weatherwise*. August/September 1993: 32–34.

Sack, Kevin. "Storm's Rains Bring Flooding in Two States." *The New York Times*. 8 September 1996: 1, 36.

Salopek, Paul. "Energy Dream Is Blowing In Midwest: Price Is Right, and Resource Is Unlimited." *The Chicago Tribune*. 6 February 1997: 1.

Schlatter, Thomas. "Weather Queries: Anatomy of a Heat Burst." *Weatherwise*. August/September 1995: 42–43.

Schlatter, Thomas. "Weather Queries: Dark Rays." *Weatherwise*. June/July 1996: 35–36.

Schlatter, Thomas. "Weather Queries: Snowrollers." *Weatherwise*. December 1996/January 1997: 42.

Sources

Shacham, Mordechai. "Danger by the Numbers: Meaningful Cold Weather Indicators." *Weatherwise*. October/November 1995: 27–28.

Shibley, John. "Glows Bands & Curtains." *Astronomy*. April 1995: 76–81.

Stevens, William K. "'95 Is Hottest Year on Record As the Global Trend Resumes." *The New York Times*. 4 January 1996: A1+.

"Stormy Weather Is On the Rise, Researchers Say Number of Blizzards, Rainstorms Jumps 20% in the U.S. Since 1900." *The Chicago Tribune*. 23 January 1997: 8.

Suplee, Curt. "Climatology: Carbon Dioxide Signals Change." *The Washington Post*. 23 September 1996: A2.

U.S. Department of Commerce. *Flash Floods*. Washington D.C.: U.S. Government Printing Office, 1982.

U.S. Department of Commerce. *Flash Floods and Floods. . . The Awesome Power!* Washington D.C.: NOAA, 1992.

U.S. Department of Commerce. *Heat Wave: A Major Summer Killer.* Washington D.C.: NOAA.

U.S. Department of Commerce. *Hurricanes. . . The Greatest Storms on Earth.* Washington D.C.: NOAA, 1994.

U.S. Department of Commerce. *NOAA Weather Radio*. Washington D.C.: NOAA, 1995.

U.S. Department of Commerce. *Thunderstorms and Lightning. . . The Underrated Killers!* Washington D.C.: NOAA, 1994.

U.S. Department of Commerce. *Tornado Safety Rules in Schools*. Washington D.C.: NOAA, 1981.

U.S. Department of Commerce. *Tornadoes. . . Nature's Most Violent Storms.* Washington D.C.: NOAA, 1992.

U.S. Department of Commerce. *Watch Out... Storms Ahead! Owlie Skywarn's Weather Book*. Washington D.C.: NOAA, 1984.

U.S. Department of Commerce. *Winter Storms: Terms to Know/How to Survive*. Washington D.C.: NOAA, 1982.

"Volunteers Pouring In to Assist Communities Ravaged by Floods." *The Los Angeles Times*. 11 March 1997: A13.

Wallace, John M. and Shawna Vogel. "El Niño and Climate Prediction." *Reports to the Nation on Our Changing Planet*. Washington D.C.: NOAA, Spring 1994.

Walsh, Edward. "Unpredictable Nature Inundates Many Towns: Flooding Hits Some Hard, Spares Others." *The Washington Post*. 7 March 1997: A3.

Walsh, Edward. "With Flood Waters At Their Door, A Few Stubborn Souls Ride It Out." *The Washington Post*. 8 March 1997: A3.

Wiche, Sandra. "Year of Extremes." *Weatherwise*. October/November 1995: 30–34.

Williams, Jack. "The Making of the Weather Page." *Weatherwise*. August/September 1992: 12–18.

Williams, Jack. "Watching the Vapor Channel: Satellites Put Forecasters On the Trail of Weather's Hidden Ingredient." *Weatherwise*. August/September 1993: 26–30.

Williams, Richard. "The Mystery of Disappearing Heat." *Weatherwise*. August/September 1996: 28–29.

Websites

Note: the following website addresses are subject to change

Anyanwu, Azunna E. O. End-to-End Forecast Process and Event-Driven Versus Schedule-Driven Products. [Online] Available http://www.nws.noaa.gov/om/etoefp.htm, November 21, 1996.

Sources

Babel Fish Corporation. Chinook Winds. *The Alberta Traveller*. [Online] Available http:www.babelfish.com/AB_Travel/weather_guides/chinook.html, January 28, 1997.

Boulder Wind Info. [Online] Available http://cdc.noaa.gov/~cas/wind.html, January 28, 1997.

Dept. of Atmospheric Sciences, Univ. of Illinois at Urbana-Champaign. Cloud Catalog. [Online] Available http://covis.atmos.uiuc.edu/guide/clouds/html/cloud.home.html, January 9, 1997.

The GOES Project. [Online] Available http://climate.gsfc.nasa.gov/~chesters/goesproject.html, November 5, 1996.

High-Resolution Weather Forecasting Readied for Olympics. *Science & Engineering News*. [Online] Available http://ike.engr.washington.edu/news/bulletin/weather.html, November 21, 1996.

Janice Huff. *INTELLiCast biographies*. [Online] Available http://www.intellicast.com/weather/bio/wnbc/jh/bio.html, November 25, 1996.

MacDonald, Michael. WeatherNet. [Online] Available http://cirrus.sprl.umich.edu/wxnet/, September 10, 1996.

National Climatic Data Center. Billion Dollar U.S. Weather Disasters, 1980–1996. [Online] Available http://www.ncdc.noaa.gov/, November 19, 1996.

National Oceanic and Atmospheric Administration. [Online] Available http://www.noaa.gov/, November 19, 1996.

National Research Council. New Radar System Aids Weather Forecasting Nationwide but May Provide Less Radar Coverage for Some Areas. [Online] Available http://xerxes.nas.edu/onpi/pr/radar/, November 6, 1996.

NEXRAD NOW. [Online] Available http://www.osf.uoknor.edu/news/vol1is1.htm, November 6, 1996.

NOAA's Geostationary and Polar-Orbiting Weather Satellites. [Online] Available http://140.90.207.25:8080/EBB/ml/genlsatl.html, November 7, 1996.

Noel, James J. More Heavy Rain Has Hit Southern Indiana and Kentucky. [Online] Available http://www.nws.noaa.gov/er/iln/afos/CRWHMDCIN, March 18, 1997.

Null, Jan. NWS Glossary. [Online] Available http://www.nws.mbay.net:80/guide.html, January 29, 1997.

Palmer, Chad. Three Roads Will Improve Forecasts. *USA Today Weather*. [Online] Available http://www.usatoday.com/weather/wforkey.htm, November 21, 1996.

Plymouth State College, New Hampshire. PSC Meteorology Program Cloud Boutique. [Online] Available http://vortex.plymouth.edu/cloud.html, January 9, 1997.

S. California Winds May Fan Wildfires. *The Salt Lake Tribune*. [Online] Available http://www.sltrib.com:80/96/OCT/25/twr/00411040.htm, January 29, 1997.

San Diego NWS. Santa Ana Winds. [Online] Available http://nimbo.wrh.noaa.gov:80/Sandiego/snawind.html, January 29, 1997.

Songer, Nancy Butler. Global Exchange Weather Program. *Kids as Global Scientists*. [Online] Available http://www-kgs.colorado.edu/index.html, November 5, 1996.

Storm Chaser Homepage. [Online] Available http://taiga.geog.niu.edu/chaser/chaser.html, November 18, 1996.

Tornado Outbreak, Flood Index. *USA Today Weather*. [Online] Available http://www.usatoday.com/weather/wlead.htm, March 18, 1997.

The Tornado Project Online. [Online] Available http://www.tornadoproject.com, November 18, 1996.

Sources

USA Today Weather. [Online] Available http://www.usatoday.com/weather/wfront.htm, November 18, 1996.

The Weather Channel. [Online] Available http://www.weather.com/, September 19, 1996.

Weather Underground. [Online] Available http://www.wunderground.com/, September 10, 1996.

WeatherNet: WeatherSites. [Online] Available http://cirrus.sprl.umich.edu/wxnet/servers.html, September 10, 1996.

Women in Weather... Mini-Biographies. [Online] Available http://www.nssl.uoknor.edu/~nws/women/biograph.html, September 10, 1996.

WX-ACCESS One: World Wide Web Amateur Weather Connection. *American Weather Observer.* [Online] Available http://members.aol.com/larrypahl/awo.htm, November 18, 1996.

CD-ROMs

Eyewitness Encyclopedia of Science. New York: Dorling Kindersley, Inc., 1994.

McGraw-Hill Multimedia Encyclopedia of Science and Technology. New York: McGraw-Hill, Inc., 1996.

The 1996 Grolier Multimedia Encyclopedia. Danbury, CT: Grolier Electronic Publishing, Inc., 1996.

Science Navigator. New York: McGraw-Hill, Inc., 1995.

Science On File CD-ROM. New York: Facts on File, Inc., 1995.

INDEX

Italic type indicates volume number;
(ill.) indicates illustration (photographs and figures).

A

Absolute humidity *1:* 46, 48-50
Absolute zero *1:* 6
Accretion *2:* 207
AccuWeather *3:* 439
Acid fog *3:* 513
Acid rain *3:* 486, 500, 512-513, 513 (ill.), 515
Acid rain experiment *3:* 515
Adiabatic processes *1:* 55
Adjustment to sea level *3:* 449-450
Advection fog *1:* 54, 112-113, 113 (ill.)
Aerogenerators *3:* 519
Aerologists *3:* 410
Agassiz, Jean Louis *3:* 480-482, 480 (ill.)
Agricultural reports *3:* 437-438, 443
Air-mass thunderstorm *2:* 218-219
Air-mass weather *1:* 34
Air masses *1:* 18; *2:* 266; *3:* 406
Air parcels *1:* 48; *2:* 212
Air pollutants *3:* 507-511, 513
Air pollution *3:* 500, 506-511, 507 (ill.), 509 (ill.)
Air pressure *1:* 13, 16, 18, 102; *3:* 422, 425, 392 (ill.), 393
Air stability. *See* Stable air layers *and* Unstable air layers
Albedo *3:* 482
Alberta Clippers *2:* 200
Alphabetical list of local winds *1:* 121
Alternative transportation *3:* 517, 518 (ill.)
Altocumulus clouds *1:* 81-82, 81 (ill.), 82 (ill.), 86, 92, 96, 99, 105, 107-108; *2:* 190
Altocumulus castellanus clouds *1:* 105, 105 (ill.)
Altocumulus undulatus clouds *1:* 96

Index

Altostratus clouds *1:* 80-83, 80 (ill.), 85-86, 88, 92, 99, 105, 107-108; *2:* 182, 199
Altostratus translucidus clouds *1:* 96
American Meteorological Society *3:* 375, 432
American Red Cross *2:* 289
Anabatic winds *1:* 124, 125 (ill.)
Anemometers *1:* 135; *2:* 248; *3:* 382, 399 (ill.)
Aneroid barometers *3:* 394, 395 (ill.)
Animals and forecasting *3:* 378-380, 379 (ill.)
Annual mean sea temperature *3:* 450 (ill.)
Annual mean temperature *3:* 449
Annual temperature range *3:* 449
Anticyclones *1:* 33, 45, 128
Arctic climates *3:* 455, 472-474
Arctic sea smoke *1:* 116
Asteroids *3:* 486, 491
Aurora australis *2:* 337-338
Aurora borealis *2:* 317, 337-338, 337 (ill.)
Auroras *2:* 337-338
Austrus *1:* 121, 130
Autumnal equinox *1:* 2
Avalanches *2:* 201-204
Aviation reports *3:* 443
Avogadro, Amedeo *1:* 15
Azores-Bermuda High *2:* 273
Azote *1:* 11

B

Backing winds *3:* 381
Ball lightning *2:* 235
Banner clouds *1:* 101, 101 (ill.)
Baquiros *2:* 264
Barchan dunes *1:* 144, 144 (ill.)
Barihs. *See* Shamals
Barographs *3:* 394, 396 (ill.)
Barometer experiment *3:* 398
Barometer readings *3:* 404 (ill.)
Barometers *1:* 16-17; *3:* 382, 394
Bead lightning *2:* 236
Beaufort, Sir Francis *3:* 400
Beaufort Wind Scale *3:* 400-401, 401 (ill.)
Bentley, William *2:* 195
Bergen School *1:* 38
Bergs *1:* 121, 136
Bises *1:* 121, 138
Bjerknes, Vilhelm *1:* 38
Black, Joseph *1:* 8, 8 (ill.)

Index

Blizzard of 1888, *2:* 205, 205 (ill.)
Blizzards *2:* 192, 200-202
Blocking highs *2:* 293, 313
Blocking lows *2:* 293
Blocking systems *2:* 292-293, 292 (ill.)
Blowing snow *2:* 200
Blue northers *1:* 121, 137
Bluestein, Howard *2:* 256-257, 256 (ill.)
Bolide winters *3:* 491
Bolides *3:* 491, 493
Boras *1:* 121, 126
Boreal. *See* Subpolar climates
Boundary layers *1:* 9
Boyle, Robert *1:* 16, 16 (ill.)
Boyle's Law *1:* 16
Breeze experiment *1:* 127, 127 (ill.)
Brick fielders *1:* 121, 136
Burans *1:* 121, 137
Burgas *1:* 121, 138
Buttes *1:* 142
Buys Ballot, Christoph *1:* 23
Buys Ballot Law *1:* 23
Byers, Horace R., *2:* 239

C

Calories *1:* 7
Carbon dating *3:* 495
Carbon monoxide *3:* 510
Carcinogens *3:* 511
Celsius, Anders *3:* 387
Celsius and Fahrenheit scales *3:* 387
Cenozoic Era *3:* 479, 483-485
Chaparral *1:* 132
Charles, Jacques Alexandre César *1:* 16
Charles' Law *1:* 16
Chichilis *1:* 121, 134
Chilis *1:* 122, 134
Chinooks *1:* 122, 128, 128 (ill.), 130, 132, 137; *2:* 204
Chlorofluorocarbons (CFCs), *3:* 500, 503, 515-516
Chroma key *3:* 434
Cirriform clouds *1:* 75, 86, 104, 106; *2:* 328
Cirrocumulus clouds *1:* 84-86, 85 (ill.), 92, 99, 107
Cirrostratus clouds *1:* 84-85, 91-92, 108; *2:* 265

Italic type indicates volume number; (ill.) indicates illustration (photographs and figures).

Index

Cirrus clouds *1:* 84, 84 (ill.), 87 (ill.), 91-92, 97, 99, 102-103, 107-108; *2:* 200, 218, 265
Cirrus floccus clouds *1:* 97
Cirrus spissatus clouds *1:* 108
Cirrus uncinus clouds *1:* 97, 97 (ill.)
Climate *3:* 448-449
Climate change *3:* 479, 485, 489-491, 493-494
Climates of the United States *3:* 477
Cloud base heights at various altitudes *1:* 76 (ill.)
Cloud classification systems *1:* 75
Cloud experiment *1:* 77
Cloud identification and forecasting *1:* 104
Cloud journal *1:* 107
Cloud seeding *2:* 190-191
Cloud species *1:* 92
Cloud-to-air lightning *2:* 235
Cloud-to-cloud lightning *2:* 235
Cloud-to-ground lightning *2:* 232-234, 233 (ill.)
Cloudbursts *2:* 186-187
Clouds *1:* 75-108
Coalescence *2:* 186, 195
Coastal floods *2:* 307
Cold. *See* Extreme cold
Cold air and warm air experiment *1:* 57
Cold clouds *1:* 64; *2:* 190, 195
Cold fronts *1:* 37, 105, 108, 115, 117; *2:* 218, 222-223
Cold-related illnesses *2:* 303, 305-306
Color of light *2:* 317-318
Columbia Gorge wind *1:* 126
Comet Shoemaker-Levy *3:* 493
Comets *3:* 480, 486
Compressional heating *1:* 125, 132
Compressional warming *1:* 56, 60, 128; *2:* 296
Computer forecasting models *3:* 420, 442, 446
Condensation *1:* 45, 50
Condensation nuclei *1:* 53, 62; *2:* 183
Conduction *1:* 6, 110, 112, 124; *2:* 261
Conservation of angular momentum *1:* 42; *2:* 249
Constructive interference *2:* 334
Continental drift *3:* 486-488, 487 (ill.)
Contrails *1:* 103, 103 (ill.)
Convection *1:* 5-6, 18, 24, 55, 61, 87-88, 90, 98-99, 132-133; *2:* 185, 220
Convective cells *2:* 218-219, 223
Conventional radar *3:* 412-414
Convergence *1:* 32, 42, 85; *2:* 219, 221, 224, 269
Coriolis, Gustave-Gaspard de. *See* De Coriolis, Gustave-Gaspard
Coriolis effect *1:* 20, 21 (ill.), 23, 26, 70, 72; *2:* 268-269

Coronas *2:* 334, 335 (ill.)
Coxwell, Robert *1:* 14, 14 (ill.)
Crepuscular rays *2:* 320, 320 (ill.)
Critical angle *2:* 332
Cumuliform clouds *1:* 75, 88, 99, 106; *2:* 185-186, 198-199
Cumulonimbus clouds *1:* 39, 88-89, 90 (ill.), 92, 98-99, 107, 108; *2:* 182-183, 195, 207, 211, 214-215, 219, 229, 231, 266
Cumulonimbus calvus clouds *1:* 93
Cumulonimbus incus clouds *1:* 90, 105
Cumulonimbus mammatus clouds *1:* 99, 99 (ill.)
Cumulus clouds *1:* 82, 83, 87 (ill.), 88, 92; *2:* 182-183, 191, 212, 214-215, 219, 221, 229, 267
Cumulus congestus clouds *1:* 89, 98-99; *2:* 258
Cumulus humilis clouds *1:* 97, 98 (ill.), 105
Cumulus mediocris clouds *1:* 89, 89 (ill.), 98
Cumulus stage *2:* 214-215
Cup anemometers *3:* 396, 399 (ill.)
Cyclogenesis *1:* 42, 43 (ill.)
Cyclones *1:* 28, 33, 41, 127; *2:* 243, 264

D

Dalton, John *1:* 15, 15 (ill.)
Dart leaders *2:* 233
De Coriolis, Gustave-Gaspard *1:* 20, 20 (ill.)
De Bort, Teisserenc *1:* 14
Decay stage *2:* 250
Deforestation *3:* 486, 502 (ill.)
Dehydration *2:* 297
Dendrites *2:* 196-197
Dendrochronology *3:* 498
Deposition *1:* 52, 62; *2:* 195, 198
Deposition nuclei *1:* 63
Derechos *1:* 122, 138, 140 (ill.); *2:* 211, 226, 237
Desert climates *3:* 454, 460-462
Desert pavement *1:* 143
Deserts *3:* 451, 460-462, 462 (ill.)
Destructive interference *2:* 334
Developing stage *2:* 214
Dew *1:* 50, 50 (ill.); *2:* 181
Dew and fog experiment *1:* 55
Dew point *1:* 15, 18, 49, 54-55, 58-59, 69, 75, 79, 81, 92, 109-110, 114, 117; *2:* 212, 214, 244; *3:* 392, 424, 464, 478
Dew-point temperature *3:* 389 (ill.)
Diamond dust *1:* 119

Italic type indicates volume number; (ill.) indicates illustration (photographs and figures).

Index

Diffraction 2: 333-337
Dispersion 2: 330-331, 334; 3: 381
Dissipating stage 2: 214
Divergence 1: 32, 42; 2: 214, 219, 221, 224
The Doctor 1: 122, 124
Doldrums 1: 25
Doppler, Christian 3: 413
Doppler effect 3: 413, 413 (ill.)
Doppler radar 2: 245, 248; 3: 412, 415, 446
Double rainbow 2: 333-334, 334 (ill.)
Downbursts 2: 218, 237-238
Downdrafts 2: 215, 218, 222, 225, 249, 296
Drainage winds 1: 125
Drifting snow 2: 200
Drifts 2: 200
Drizzle 2: 182-183, 185, 187, 198
Dropwindsondes 3: 409
Droughts 2: 293, 312-313, 312 (ill.), 316; 3: 459
Dry adiabatic lapse rate 1: 56, 128; 2: 214
Dry-bulb thermometers 3: 388, 390
Dry climates 3: 460
Dry tongue 2: 224
Dust Bowl 2: 313-316, 314 (ill.)
Dust storms 1: 132-135, 134 (ill.)
Dust devils 1: 122, 133, 133 (ill.)
Dust-whirl stage 2: 249

E

Earth's atmosphere 1: 9-13, 12 (ill.)
Eccentricity of Earth's orbit 3: 488-489
Eddies 1: 102
Ekman Spiral 1: 72
El Niño/Southern Oscillation (ENSO), 1: 72-74, 73 (ill.); 3: 486
El norte 1: 122, 137
Emperor penguins 3: 475, 475 (ill.)
Ensemble forecasting 3: 446-447
Entrainment 2: 216
Equinoxes 3: 474
European Satellite Agency (ESA), 3: 416
Evaporation 1: 46, 51; 2: 193; 3: 453
Evaporation fog 1: 54, 114, 116
Expansional cooling 1: 56, 117
Extratropical cyclones 1: 41; 2: 266
Extreme cold 2: 301-306, 305 (ill.)
Extreme heat 2: 293-294, 294 (ill.), 297-298
Eye wall 2: 265, 285

F

Faculae *3:* 374, 493
Fahrenheit, Gabriel Daniel *1:* 8; *3:* 387
Fahrenheit and Celsius Scales *3:* 387
Fair-weather waterspouts *2:* 257
Fall streaks *2:* 200
Fata Morgana *2:* 327, 330
Federal Emergency Management Agency (FEMA), *2:* 289
Ferrel, William *1:* 25
Ferrel cells *1:* 25
Fetch *1:* 149
Flash flood alerts and safety procedures *2:* 240
Flash flood warnings *2:* 240
Flash flood watches *2:* 240
Flash floods *2:* 191, 220-221, 226, 238-243, 306, 308, 309 (ill.)
Floods *2:* 293, 306-312, 311 (ill.); *3:* 459
Flurries *2:* 192, 199
Foehns *1:* 122, 130, 137
Fog *1:* 109-120
Fog stratus *1:* 119-120, 119 (ill.)
Forecasting *3:* 369-447
Forecasting equipment *3:* 408
Forked lightning *2:* 236
Fractocumulus clouds *1:* 98
Franklin, Benjamin *2:* 228-229, 228 (ill.)
Freezing drizzle *2:* 188
Freezing fog *1:* 118
Freezing nuclei *1:* 53, 63; *2:* 190
Freezing rain *2:* 182, 187-189, 192, 205
Frontal fog *1:* 115-117
Frontal system *1:* 104-105
Frontal thunderstorms *2:* 219-220, 223
Frontal uplift *1:* 61
Fronts *1:* 37, 55; *2:* 219, 229, 266; *3:* 429-430
Frost *2:* 181
Frost fairs *3:* 486
Frostbite *1:* 9; *2:* 201, 303, 305, 306 (ill.); *3:* 422
Fujita, Tetsuya Theodore *2:* 237-239, 238 (ill.), 253, 285
Fujita Intensity Scale *2:* 239, 252-254, 254 (ill.)
Fujita-Pearson Scale. *See* Fujita Intensity Scale
Funnel clouds *2:* 221, 244
Future of forecasting *3:* 444

Italic type indicates volume number; (ill.) indicates illustration (photographs and figures).

Index

G

Gay-Lussac, Joseph-Louis *1:* 16
Gay-Lussac's Law *1:* 16
Geostationary satellites *3:* 417-418
Gharbis *1:* 122, 135
Glaciers *3:* 480-481, 483-485, 484 (ill.), 487-488, 490, 495, 505
Glaisher, James *1:* 14, 14 (ill.)
Glaze *2:* 188-189, 192
Global warming *3:* 485, 502-505
Global water budget *1:* 67
Global wet and dry zones *3:* 452 (ill.)
Global wind patterns *1:* 24 (ill.)
Glories *2:* 336-337
GOES satellites *3:* 418-419, 419 (ill.)
The Grapes of Wrath, *2:* 315
Graupel *2:* 198, 207, 209, 231
Gravity *1:* 63
Gravity winds *1:* 125
The Great Depression *2:* 315
"Great Ice Age", *3:* 481
Green flash experiment *2:* 326
Green flashes *2:* 323-324
Greenhouse effect *3:* 493, 500-503, 501 (ill.)
Greenhouse gases *3:* 493, 500, 502-503
Greenpeace *3:* 506
Ground blizzards *2:* 200
Ground fog *1:* 110, 111 (ill.)
Ground-to-cloud lightning *2:* 234-235
Gust fronts *2:* 222, 249, 259
Gyres *1:* 71

H

Haboobs *1:* 122, 132
Hadley, George *1:* 23
Hadley cells *1:* 23, 26
Hail *2:* 181, 206, 209, 211, 223, 226, 231
Hail alley *2:* 207
Hail suppression *2:* 207-208
Hailstones *2:* 204, 206-209, 206 (ill.)
Hailstorms *2:* 207, 220
Hair hygrometers *3:* 392-393
Haloes *2:* 328-329, 329 (ill.), 331, 334; *3:* 381
Harmattans *1:* 122, 133
Haze *2:* 319-320
Health risks of extreme cold *2:* 301-306

Index

Health risks of extreme heat *2:* 293-294, 297-298
Heat. *See* Extreme heat
Heat bursts *2:* 296
Heat cramps *2:* 299
Heat equator *1:* 24
Heat exhaustion *2:* 297, 299
Heat lightning *2:* 236
Heat-related illnesses *2:* 297-300
Heat stroke *2:* 297-300
Heat syncope *2:* 299
Heat waves *2:* 297, 300-301
Heating-degree-days *3:* 423, 434
Heavy snow *2:* 199
Height of cloud bases *1:* 91 (ill.)
High fog *1:* 119
Hinrichs, Gustav *1:* 138
History of climate change *3:* 479
Hoar frost *1:* 52, 52 (ill.), 118
Hollow-column snowflakes *2:* 196-197
Holocene epoch *3:* 483
Horse latitudes *1:* 25; *3:* 461
Howard, Luke *1:* 75
Huff, Janice *3:* 435, 435 (ill.)
Human activity and the future *3:* 500-520
Humid subtropical climates *3:* 454, 464-465
Humid tropical climates *3:* 455
Humidity *1:* 39; *3:* 387, 393, 422-423
Humiture index *3:* 424, 425 (ill.)
Huracans *2:* 264
Hurricane, *2:* 279 (ill.)
Hurricane Agnes *2:* 267
Hurricane Allen *2:* 287
Hurricane Andrew *2:* 282, 282 (ill.)
Hurricane Camille *2:* 284, 287
Hurricane Diane *2:* 284
Hurricane Elena *2:* 278
Hurricane Gilbert *2:* 276, 277 (ill.)
Hurricane Hugo *2:* 283
Hurricane Iniki *2:* 275, 275 (ill.)
Hurricane names *2:* 289-290, 290 (ill.)
Hurricane research jet *3:* 412
Hurricane warnings *2:* 288-289
Hurricane watches *2:* 288-289
Hurricanes *2:* 226, 241, 243, 262-290, 263 (ill.), 264 (ill.), 270 (ill.), 273 (ill.), 274 (ill.), 307; *3:* 412

Italic type indicates volume number; (ill.) indicates illustration (photographs and figures).

Index

Hygrometers *3:* 382
Hypothermia *2:* 201, 303, 305

I

Ice *2:* 181, 204
Ice ages *3:* 479, 482, 505
Ice cores *3:* 496-497
Ice fog *1:* 118-119
Ice pellets *2:* 181, 204
Ice storms *2:* 188-189, 188 (ill.)
Industrial Revolution *3:* 503
Inferior mirages *2:* 326, 327 (ill.)
Infrared satellite image *3:* 417 (ill.)
Instrument shelters *3:* 382-383, 383 (ill.)
Inter-cloud lightning *2:* 235
International Meteorological Congress *1:* 75
International weather symbols *3:* 400, 427, 428 (ill.)
Intertropical convergence zone (ITCZ), *1:* 25; *2:* 269
Inversion *1:* 60; *2:* 188, 221, 269; *3:* 511
Iridescence *2:* 335-336
Iridescent clouds *2:* 336 (ill.)
Isobars *3:* 429, 430 (ill.), 434
Isotherms *3:* 434, 450-451

J

Jet maximum *1:* 32
Jet stream *1:* 30-31, 90, 97; *2:* 224, 267; *3:* 434

K

Katabatic winds *1:* 125-128, 125 (ill.), 148
Keller, Will *2:* 258-259
Kelvin, Baron *1:* 103
Kelvin-Helmholtz clouds *1:* 102, 102 (ill.)
Khamsins *1:* 122, 134
Kinetic energy *1:* 6, 56
Knight, Nancy *2:* 197
Konas *1:* 122, 141
Köppen, Wladimir *3:* 453
Köppen system *3:* 453
Kudloo, Tom *3:* 410

L

Lake breezes *1:* 124
Lake-effect snow *2:* 200, 201 (ill.)
Land breeze experiment *1:* 68
Land breezes *1:* 67, 113, 123, 123 (ill.), 140
Latent heat *1:* 7, 42, 51, 57-58, 67; *2:* 191, 217, 231, 259, 268; *3:* 390
Lavoisier, Antoine-Laurent *1:* 11, 11 (ill.)
Leeward slopes *1:* 69, 101-102, 127-128, 130, 142; *3:* 453, 478
Lenticular clouds *1:* 100, 100 (ill.)
Lestes *1:* 122, 134
Levanters *1:* 122, 141
Leveches *1:* 122, 134
Light, color of. *See* Color of light
Light, scattering of. *See* Scattering of light
Lightning *2:* 211, 215, 226, 229, 230 (ill.), 232-236, 243, 248
Lightning rods *2:* 229, 236, 236 (ill.)
Lightning safety *2:* 227
Little Ice Age *3:* 485-486, 494
Local winds *1:* 121-149
Lorenz, Edward *3:* 376, 377 (ill.), 444

M

Macrobursts *2:* 226, 237
Major cloud groups *1:* 77 (ill.)
Major El Niño event *1:* 73
Mammatus clouds *1:* 99; *2:* 222
Marine climates *3:* 455, 464-466
Marine forecasts *3:* 443, 445
Marine reports *3:* 438
Maritime tropical air mass *2:* 219
Mature stage *2:* 214-215, 217, 229, 231, 249
Maunder, E. W., *3:* 494
Maunder minimum *3:* 494
Maximum and minimum thermometers *3:* 385 (ill.), 386-387, 405
Measuring atmospheric conditions *3:* 385
Measuring climate change *3:* 494
Media weathercasting *3:* 431
Mediterranean climates *3:* 455, 464, 466-467, 466 (ill.)
Melting zone *2:* 189
Meltwater equivalent *3:* 398-399
Mercury barometers *3:* 393-394, 393 (ill.)
Mesocyclones *2:* 226, 248-249
Mesoscale convective complex *2:* 220

Italic type indicates volume number; (ill.) indicates illustration (photographs and figures).

Index

Mesoscale winds *1:* 121
Mesozoic Era *3:* 479, 482-483
Meteorology *1:* 31, 65, 94; *2:* 253, 257; *3:* 369, 374-375, 435, 439, 444
Methane *3:* 500, 504, 511
Microbursts *2:* 226, 237, 239; *3:* 414
Microwaves *3:* 412-413
Mid-latitude cyclones *3:* 475
Middle latitudes *1:* 18; *2:* 189, 243, 269, 273, 292-293, 301, 313, 337; *3:* 461, 464
Migrating winds *1:* 136 (ill.)
Milankovitch, Milutin *3:* 488
Milankovitch theory *3:* 489, 489 (ill.)
Minimum thermometers. *See* Maximum and minimum thermometers
Mirages *2:* 317, 324-327, 325 (ill.)
Mist *1:* 109
Mistrals *1:* 122, 126
Moist adiabatic lapse rate *1:* 56-58; *2:* 215
Monsoon climates *3:* 454, 457-458
Monsoons *1:* 140, 142 (ill.); *3:* 448, 456-459
Moon dogs *2:* 330
Motor vehicle emissions *3:* 508, 511
Mountain breezes *1:* 124-125
Mountain climates *3:* 474, 476, 476 (ill.)
Mountain thunderstorms *2:* 219
Mountain-wave clouds *1:* 100-101, 104 (ill.); *2:* 185, 185 (ill.)
Mountain weather *1:* 69 (ill.)
Muir glacier *3:* 484 (ill.)
Multi-vortex tornadoes *2:* 249, 250 (ill.)
Multicell thunderstorms *2:* 218

N

National Aeronautics and Space Administration (NASA), *3:* 418, 504, 512
National Center for Atmospheric Research *2:* 197
National Hurricane Center *2:* 286, 288
National Meteorological Center *3:* 420
National Oceanic and Atmospheric Administration (NOAA), *3:* 372, 412, 418, 437
National Severe Storm Forecast Center *2:* 253
National Weather Association *3:* 432
National Weather Service (NWS), *1:* 130, 135; *2:* 200, 202, 221, 288-289; *3:* 372, 382, 397, 399, 408, 414, 420, 433, 438-439, 443, 446-447
National Weather Service forecast office *3:* 373 (ill.)
Natural weather phenomena *3:* 377
Newspaper weathercasts *3:* 433, 437
Newspaper weather map *3:* 436 (ill.)
NEXRAD radar system *3:* 414, 414 (ill.)
Nimbostratus clouds *1:* 75, 78, 80-81, 88, 92, 96, 105, 107-108; *2:* 182, 199
Nimbus clouds *1:* 75

Index

Nitric acid *3:* 512
Nitric oxide *3:* 511
Nitrogen dioxide *3:* 511-512
Nitrogen oxides *3:* 511
NOAA Weather Radio *3:* 437-438
NOAA-K satellites *3:* 419
Non-tornadic waterspouts *2:* 257
Nor'easter *1:* 122, 137
Northeaster *1:* 137
Northern temperate climates *3:* 455, 467-469, 468 (ill.), 476-477
Nor'wester *1:* 122, 130

O

Obliquity *3:* 490
Occluded fronts *1:* 40, 40 (ill.), 44
Ocean currents *1:* 22, 70, 71 (ill.); *3:* 451, 488
Ocean and lake sediments *3:* 497
Oceanographic satellites *3:* 419
Optical effects *2:* 317-338
Organized convection theory *2:* 270
Organizing stage *2:* 249
Orographic clouds *1:* 100
Orographic lifting *1:* 61, 68; *3:* 453
Orographic thunderstorms *2:* 219
Overshooting *2:* 222
Ozone depletion *3:* 500, 513-516
Ozone hole *3:* 514
Ozone layer *1:* 4, 11, 13; *3:* 514-516

P

Paleoclimatologists *3:* 494
Paleozoic Era *3:* 479, 482
Pamperos *1:* 122, 138
Pangea *3:* 482
Papagayos *1:* 122, 126
Paper hygrometer experiment *3:* 390-391
Particulate matter *3:* 509-510
Pascal, Blaise *1:* 17, 17 (ill.)
Permafrost *3:* 470, 472
Phase changes *1:* 7
Phased array antennas *3:* 415
Photochemical smog *3:* 509 (ill.)

Italic type indicates volume number; (ill.) indicates illustration (photographs and figures).

Index

Photovoltaic cells *3:* 518
Pileus *1:* 99
Plant water loss experiment *3:* 463
Plants and forecasting *3:* 380, 380 (ill.)
Polar air masses *3:* 407
Polar bear *3:* 473 (ill.)
Polar cells *1:* 26
Polar climates *3:* 470-471, 473 (ill.)
Polar easterlies *1:* 26, 28, 31, 121; *3:* 451
Polar fronts *1:* 26, 31; *3:* 375
Polar-orbiting satellite *3:* 417, 418 (ill.)
Polar region *3:* 451
Pollution *3:* 500, 506-511, 507 (ill.), 509 (ill.), 510 (ill.), 513
Poseidon satellite *3:* 420
Power plant pollution *3:* 510 (ill.)
Prairies *3:* 463
Pre-Holocene epoch *3:* 484-485
Precambrian Era *3:* 479, 494
Precession of the equinoxes *3:* 490
Precipitation *2:* 181-210, 182 (ill.); *3:* 422, 426, 451, 453
Precipitation fog *1:* 116
Precipitation measurement *3:* 396-398
Pressure gradient force (PGF), *1:* 19
Primary air pollutants *3:* 509
Primary rainbows *2:* 333
Psychrometers *3:* 382, 388
Purgas *1:* 122, 138

R

Radar *2:* 237, 288; *3:* 412
Radar waves *3:* 413
Radiation fog *1:* 54, 109, 110 (ill.)
Radiational cooling *1:* 110, 124; *2:* 188
Radio weathercasts *3:* 437
Radioactive dating *3:* 494-495
Radiometers *3:* 493
Radiosondes *3:* 371-372, 409, 409 (ill.)
Rain *2:* 181-183, 183 (ill.), 187
Rain bands *2:* 265
Rain gauges *3:* 382, 397-398, 402 (ill.)
Rain-shadow effect *1:* 69
Rain shadows *3:* 478
Rainbow experiment *2:* 338
Rainbows *2:* 331, 333 (ill.), 334; *3:* 381
Rainfall *2:* 187
Rainforest climates *3:* 451, 454, 456-457, 456 (ill.)

Index

Rainmaking *2:* 190-191, 208
The Rains Came, *3:* 458 (ill.)
Rawinsondes *3:* 409, 415
Recording barographs *3:* 397 (ill.)
Recording weather observations *3:* 399-400
Recycling *3:* 517
Reduction to sea level *3:* 393
Reflection *2:* 318, 331, 336
Refraction *2:* 318, 322 (ill.), 324 (ill.), 331, 336; *3:* 381
Refraction experiment *2:* 323
Regional winds *1:* 121
Relative humidity *1:* 15, 48-50, 49 (ill.); *2:* 193, 216; *3:* 388, 388 (ill.), 392, 399, 423
Return strokes *2:* 232
Ribbon lightning *2:* 236
Richardson, Lewis Fry *3:* 421
Ridges *1:* 28; *2:* 224
Rime *1:* 118, 118 (ill.); *2:* 192, 209
Riming *2:* 198
River floods *2:* 307-308
Rocks and rock formations *3:* 494-496
Roll clouds *2:* 222
Rossby, Carl-Gustaf *1:* 30, 31; *3:* 374
Rossby waves *1:* 30-31
Rutherford, Daniel *1:* 11

S

Saffir, Herbert *2:* 286
Saffir-Simpson Hurricane Intensity Scale *2:* 286, 287 (ill.)
Saguaro National Monument desert *3:* 462 (ill.)
Saltation *1:* 143, 143 (ill.)
Salting roads *2:* 194
Sample weather log *3:* 402
Sand dunes *1:* 143 (ill.), 144
Sand ripples *1:* 145, 146
Sandstorm *2:* 222, 229
Santa Ana winds *1:* 122, 130, 130 (ill.), 137
Sastrugis *1:* 146
Saturation point *1:* 15, 47, 54, 109, 114, 117
Savanna climates *3:* 454, 459, 459 (ill.)
Savannas *3:* 456, 459, 459 (ill.)
Scattering of light *2:* 318-319
Schaefer, Vincent *2:* 190
Sea breeze experiment *1:* 68

Italic type indicates volume number; (ill.) indicates illustration (photographs and figures).

Index

Sea breezes *1:* 67, 113, 123-124, 123 (ill.), 131, 140; *2:* 258
Sea fog *1:* 113-114
Seasat satellites *3:* 419
Seasons *1:* 1, 2 (ill.)
Seasons experiment *1:* 3
Secondary air pollutants *3:* 509
Secondary rainbows *2:* 333
Sector plates *2:* 196
Seif dunes *1:* 145
Selective scattering *2:* 318
Semipermanent highs and lows *1:* 27, *2:* 291; *3:* 475
Severe blizzards *2:* 201
Severe thunderstorms *2:* 211, 221
Severe weather warnings *3:* 441-442
Shahalis *1:* 122, 134
Shamals *1:* 122, 136
Sharavs *1:* 122, 135
Sheet lightning *2:* 235
Shelf clouds *2:* 222
Short waves *1:* 30
Shrinking stage *2:* 250
Simooms *1:* 122, 135
Simpson, Robert *2:* 286
Single-cell thunderstorms *2:* 218
Sinkholes *2:* 308
Siroccos *1:* 122, 134, 137
Ski reports *3:* 444
Skilled forecasts *3:* 370-371
Sleet *1:* 138; *2:* 181, 206
Sling psychrometers *3:* 392
Smog *3:* 486, 507-508, 508 (ill.)
Snow *2:* 181, 189, 192-193, 193 (ill.), 195-196, 195 (ill.), 196 (ill.), 198, 199 (ill.), 200
Snow dunes *1:* 146
Snow fences *1:* 147, 147 (ill.)
Snow grains *2:* 198
Snow pellets *2:* 198
Snow ripples *1:* 146
Snow squalls *2:* 199
Snowflakes *2:* 192, 195, 195 (ill.), 196 (ill.)
Snowrollers *2:* 204
Soft hail *2:* 198
Solar cars *3:* 518, 518 (ill.)
Solar energy *1:* 1, 3; *3:* 517-519
Solar variability *3:* 493
Soundings *3:* 417
Southerly buster *1:* 122, 137
Southern Oscillation. *See* El Niño/Southern Oscillation (ENSO)

Index

Specialized forecasts *3:* 443, 444
Specific heat *1:* 7-8
Spin-up vortices *2:* 285
Spontaneous nucleation *1:* 63
Squall lines *1:* 39, 90, 106, 139; *2:* 220, 223-224, 247
Stable air layers *1:* 58 (ill.), 59, 83; *2:* 212, 215
Stable atmosphere. *See* Stable air layers
Stationary fronts *2:* 310
Steam devils *1:* 116
Steam fog *1:* 115-116, 115 (ill.); *2:* 200
Steppe climates *3:* 454, 462-463, 464 (ill.)
Stepped leaders *2:* 232-235
Stevenson screen *3:* 382
Storm surges *2:* 278-281, 280 (ill.), 286, 307
Storm tides *2:* 279
Storm-tracking image *1:* 44 (ill.)
Stratiform clouds *1:* 75
Stratocumulus clouds *1:* 75, 78 (ill.), 82, 82 (ill.), 88, 92, 96, 98-99, 105, 107-108
Stratocumulus stratiformis clouds *1:* 96
Stratus clouds *1:* 78, 79, 79 (ill.), 92, 96, 105, 108; *2:* 183, 198, 229
Stratus fractus clouds *1:* 96
Studies on Glaciers, *3:* 481
Sublimation *1:* 52; *2:* 200
Subpolar climates *3:* 455, 467, 469-470, 499
Subsidence *2:* 308
Subtropical climates *3:* 464, 467
Suction vortices *2:* 249
Sulfur dioxide *3:* 510, 512
Sulfuric acid *3:* 512
Sundogs *2:* 329-330, 331 (ill.)
Sunspots *3:* 374, 493-494
Supercell thunderstorms *2:* 223-225, 224 (ill.), 225 (ill.), 245, 248, 257
Supercooled water *1:* 52-53, 63-65, 118; *2:* 181, 188, 190, 192, 208-209
Superior mirages *2:* 327, 328 (ill.)

T

Taiga. *See* Subpolar climates
Tehuantepecers *1:* 122, 138
Television weathercasting *3:* 431, 432 (ill.)
Temperate climates *3:* 455, 467-469, 468 (ill.), 476-477
Temperate latitudes *1:* 84
Temperature extremes *2:* 291-306
Temperature inversions *3:* 511

Italic type indicates volume number; (ill.) indicates illustration (photographs and figures).

Index

Ten deadliest U.S. hurricanes *2:* 283
Terminal velocity *1:* 63, 64
Texas northers *1:* 122, 136, 138
Thermals *2:* 214, 219
Thermographs *3:* 384 (ill.), 385
Thermometers *3:* 382-383, 385-388, 385 (ill.), 406
Thornthwaite, C. Warren *3:* 453-454
Thornthwaite classification *3:* 453
Thunder *2:* 215, 226-227, 234-236, 242
Thunderstorms *2:* 207, 211-242, 213 (ill.), 217 (ill.), 224 (ill.), 225 (ill.)
TIROS satellites *3:* 416, 416 (ill.), 419
TOPEX satellites *3:* 420
Tornadic waterspouts *2:* 256
Tornado alerts and safety procedures *2:* 260-261
Tornado cyclones *2:* 249
Tornado disasters *2:* 246
Tornado families *2:* 245
Tornado warnings *2:* 260
Tornado watches *2:* 260
Tornadoes *2:* 211, 217, 221, 223, 226, 243-261, 244 (ill.), 247 (ill.), 248 (ill.), 251 (ill.), 252 (ill.), 276, 285
Torricelli, Evangelista *3:* 392-393, 392 (ill.)
Towering cumulus clouds *1:* 86 (ill.); *2:* 183, 214
Trade winds *1:* 24, 28, 31, 121, 141; *2:* 261, 273; *3:* 451, 457
Transpiration *1:* 66; *3:* 453
Transverse dunes *1:* 145
Traveler's reports *3:* 443
Tree rings *3:* 497 (ill.), 498-499
Tropical air masses *3:* 407
Tropical cyclones *1:* 41; *2:* 263
Tropical depressions *2:* 263, 272
Tropical disturbances *2:* 263, 268, 271
Tropical Prediction Center *2:* 288
Tropical rainforests *3:* 451, 454, 456-457, 456 (ill.)
Tropical squall cluster *2:* 267
Tropical storms *2:* 263, 268, 272, 281-282, 287, 289
Troughs *1:* 28, 127; *2:* 224, 269
Tsunamis *1:* 149; *2:* 308
Tundra climates *3:* 455, 471
Tundras *3:* 471
Twister, *2:* 252, 256
Twisters *2:* 243
Typhoons *2:* 264

U

United Nations *3:* 505

Index

Unskilled forecasts *3:* 370-371
Unstable air layers *1:* 59, 59 (ill.), 83; *2:* 212, 220
Unstable atmosphere. *See* Unstable air layers
Updrafts *2:* 183, 186, 207, 216, 218, 221-222, 225-226, 248, 259
Upper-air westerlies *1:* 26, 29, 29 (ill.), 31; *2:* 224, 292
Upslope fog *1:* 116 (ill.), 117
Upwelling *1:* 71-72; *3:* 451, 466
U.S. Weather Bureau *2:* 281

V

Valley breezes *1:* 124
Valley fog *1:* 111, 112 (ill.), 119
Vapor channel *3:* 417
Vapor pressure *1:* 64; *2:* 197
Veering winds *3:* 381
Ventifacts *1:* 145
Vernal equinoxes *1:* 2
Vertical clouds *1:* 124
Virga *1:* 86, 105; *2:* 185, 186 (ill.), 200, 237; *3:* 460
Visible radiation *3:* 416-417
Volatile organic compounds (VOCs), *3:* 511
Volcanoes *3:* 486, 490-492, 492 (ill.)
Von Helmholtz, Hermann *1:* 103
Vonnegut, Bernard *2:* 190
Vortexes *2:* 243

W

Wall clouds *2:* 255
Warm clouds *1:* 66; *2:* 182
Warm fronts *1:* 37, 85-86, 105-108, 115, 117; *2:* 220
Water cycles *1:* 66; *2:* 190
Water mirages *2:* 325
Waterspouts *2:* 255, 257-261
Weather aircraft *2:* 288; *3:* 410-412, 411 (ill.)
Weather balloons *3:* 409
Weather Central *3:* 439
The Weather Channel *3:* 432
Weather forecasts *3:* 422-427, 441-442
Weather instrument resources *3:* 383-384
Weather log *3:* 402
Weather maps *3:* 428-430, 431 (ill.)
Weather modification *2:* 190

Italic type indicates volume number; (ill.) indicates illustration (photographs and figures).

Index

Weather satellites *3:* 415-416, 418, 442, 446
Weather station entry *3:* 428-429, 429 (ill.)
Weather websites *3:* 439-441
WeatherData *3:* 439
Wegener, Alfred *1:* 65, 65 (ill.); *3:* 486
Westerlies *1:* 26, 28, 30-31, 121; *2:* 273; *3:* 451
Wet-bulb depression *3:* 392
Wet-bulb temperature *3:* 424
Wet-bulb thermometers *2:* 193; *3:* 388, 390, 424
Whirlwinds *1:* 122, 133
Whirlys *1:* 122, 138
Whiteouts *2:* 201-202
Willett, Hurd *3:* 374-375, 374 (ill.)
Willy-nillys *1:* 122, 133
Willy-willies *2:* 264
Wind direction measurement *3:* 394-395
Wind energy *3:* 519-520
Wind farms *3:* 519-520, 520 (ill.)
Wind profilers *3:* 415, 446
Wind shear *1:* 102, 133; *2:* 225-226, 226 (ill.), 237; *3:* 414
Wind socks *3:* 382, 394-396
Wind speed measurement *3:* 396
Wind turbines *3:* 519
Wind vanes *3:* 394, 396, 399 (ill.)
Wind waves *1:* 147; *2:* 308
Windchill *1:* 9-10
Windchill equivalent temperature (WET), *1:* 9-10; *2:* 304; *3:* 422
Windiest places in the world *1:* 148-149
Windmills *3:* 519, 519 (ill.)
Winds acting on a hot air balloon *1:* 22 (ill.)
Winds aloft *1:* 28; *2:* 214, 219, 221, 224-225, 239, 269; *3:* 415
Windward slopes *1:* 69, 101-102, 128, 130; *3:* 453, 478
Winter storm alerts and safety procedures *2:* 202-203
Winter storm warnings *2:* 202
Winter storm watches *2:* 202
Winter weather advisories *2:* 202
Wooly Lamb *3:* 433
World cold records *2:* 301-304
World heat records *2:* 294-295, 297
World Meteorological Centers *3:* 372
World Meteorological Organization (WMO), *2:* 290; *3:* 371
World Weather Watch *3:* 371

Z

Zondas *1:* 122, 130

NORKAM SECONDARY SCHOOL
730 - 12th STREET
KAMLOOPS, BC V2B 3C1
PH (250) 376-1272
FAX (250) 376-3142